THE PERFECT BABY

The Perfect Baby

A Pragmatic Approach to Genetics

Glenn McGee

ROWMAN & LITTLEFIELD PUBLISHERS, INC.
Lanham • Boulder • New York • London

ROWMAN & LITTLEFIELD PUBLISHERS, INC.

Published in the United States of America
by Rowman & Littlefield Publishers, Inc.
4720 Boston Way, Lanham, Maryland 20706

3 Henrietta Street
London WC2E 8LU, England

British Cataloging in Publication Information Available

Library of Congress Cataloging-in-Publication Data

McGee, Glenn, 1967–
The perfect baby: a pragmatic approach to genetics / Glenn McGee.
p. cm.
Includes bibliographical references and index.
ISBN 0-8476-8343-5 (cloth: alk. paper).—ISBN 0-8476-8344-3 (paper: alk. paper)
1. Eugenics. 2. Genetic engineering. I. Title.
HQ751.M358 1997
304.5—dc20 96-31064
CIP

ISBN 0-8476-8343-5 (cloth : alk. paper)
ISBN 0-8476-8344-3 (pbk. : alk. paper)

Printed in the United States of America

™ The paper used in this publication meets the minimum requirements of American National Standard for Information Sciences—Permanence of Paper for Printed Library Materials, ANSI Z39.48-1984.

Contents

Preface

We say we have found the perfect wine for a particular dish. The perfect bride is known so only by her groom, and celebrated as such only for a week. The perfect day brings together elements of atmosphere, emotion, and planning. We know it when we experience it, and we plan it with breathless anticipation and the recognition that rain means doom. What is the *perfect baby*? Parents tell us that whichever baby they bring into the world is perfect. And we smile and acknowledge that the making (and having) of a baby is indeed an epiphany unlike any other we are likely to experience. We celebrate as a perfection that time in which some strange constellation of luck, planning, and biology sweeps over us like a (perfect?) wave. Perfect babies can have big ears and wail like banshees, they may be blind or have fewer fingers than other babies.

But the perfect wine is also an advertising campaign, a way of promising that some brand of wine is better than the others. And the perfect bride is sketched out in gruesome detail in the wedding magazines, an icon to which virtually no one lives up. The perfect day is a picture that sells beer and cigarettes. And the perfect baby is becoming a subtle commercialization of the same ideal traits that shaped eugenics at the turn of the century. Biotechnology companies rush ahead full speed to develop genetic tests that will tell families very little, but allow very grave choices. How are families to decide whether or not to abort a child who might sixty years later develop Alzheimer's disease, who might forty-five years later go mad from Huntington's chorea, who might twenty years later die of cystic fibrosis or breast cancer, or who might manifest Down's syndrome in a couple of years? Genetic counselors are so acutely aware of the historical tragedies associated with eugenics that they have adopted strange and hollow neutrality. Legally, today it is every woman for herself. Morally, no institution of society has acknowledged our collective responsibility to think about and develop the wisdom and methods for new decisions about genetic research, genetic tests, and gene therapies.

Almost every day, the "magic code" for another disease is unveiled in the pages of the major dailies. Strangely, though, the discussion of genetic tests and gene therapy that we find in the media and popular press seems alarmist and misinformed. Not surprisingly, while academic conversation among those working in philosophy and bioethics is more sophisticated, it is all but out of reach of the layperson and disconnected from the average choices we all must make about parenthood—and that we must make as a society about allocating our health care resources.

There is a parable about a drunken man who loses his wallet. A policeman sees him fumbling beneath a bright streetlamp and asks him what he's doing. The man says, "I lost my wallet over there," pointing to a dark place some thirty feet away. The policeman looks puzzled: "If you lost it over there, why are you searching for it over here?" The drunken man doesn't pause: "Here is where the light is!" During much of my training as a philosopher I felt just this way. So much of our respected and published philosophy eschews the murky, dark, and smelly places where human life happens. When we have a problem, instead of hunting for answers in the dangerous contexts of real life, we go to the light. Abstract thought, which is our great disciplinary strength, can also be our Achilles' heel.

This book is an attempt to think and write about genetic interventions in a pragmatic way. Rather than reviewing the attacks on or defenses of genetic testing that have been written by the most prominent philosophers and theologians, we deal here with the texts actually read by scientists and parents. And we think about the ideas that are in play in our actual discussion of parenthood and babies, about ideas like identity, perfection, enhancement, and illness. We return to the philosophical method developed by William James and John Dewey, in the first book-length attempt to use classical American pragmatism to solve a problem in bioethics. It is an important method, one that brings the importance of empirical investigation and scientific discovery to bear on moral theory, and brings moral theory to the actual applications where it must matter.

The history of genetics in the past two hundred years could in some ways be characterized as a race toward the Human Genome Project. The project's goals, ambitious and variegated, are united by a common faith that hereditary information will be instrumental in the development of medical technologies. While the technologies produced by the project are new, our cultural and social goals for the use of those technologies are rooted in the history of human inquiry into heredity. In science and in families, there is a long history of interest in improving the quality of offspring. This history is, in some ways, infamous: societies have sterilized hundreds of thousands of people in the interest of eugenics. But decisions about the purposes and context of reproduction are unavoidable, and are always made in a social context. People talk about perfect babies, whether or not they use a blueprint to define them.

As the new reproductive and gene therapy technologies make diagnosis and cure of genetic anomalies possible, long-standing questions about social control of reproduction will have to be confronted. Hopes and fears concerning human genetic exploration have been catalyzed by the Human Genome Project. Many believe that genetic engineering will radically alter human experience in wonderful, dangerous, or disastrous ways. In this book we will think about some intelligent goals for genetics within its social and political context.

In Chapter 1, we take a look at the new landscape of genetic technologies, practical, personal, and scientific. Then we examine what the prominent members of the public discussion of genetics are saying. In Chapter 2, the *hopes* of several biologists and contemporary social critics are arrayed. In Chapter 3, those who *fear* human genetic engineering are presented. We will try to square these hopes and fears with the reality of parenting in an era of genetics.

In Chapters 4 and 5, we will see that most of the hope and fear about genetics is based on misinformation. Those who seek a GenEthics have put us on the wrong track. We need practical wisdom. The pragmatic approach to ethics and genetics is based on the idea that the answers scientists need to develop research and clinical priorities, and the answers parents need in order to make difficult decisions, are found in the contexts of good science and good parenting. For the philosophical roots of this view we turn to the work of two American pragmatists, William James and John Dewey.

In Chapter 6 we put the pragmatic approach to work, and develop some practical principles for using genetics to deal with health problems. The questions of genetic testing, genetic diagnosis, genetic etiology, and genetic therapies are discussed. If you are facing a choice about genetic testing or gene therapy, we will identify some ways to think about the decision you have to make. If you are thinking about where to go with your genetic research into disease or illness, we will develop criteria to consider.

We then turn to the possibility of improving conditions through genetics. It is held in Chapter 7 that the use of genetic means to improve humanity is dangerous, but no more morally problematic than the use of piano lessons, mega vitamins, and expensive private school. We all, in the final analysis, want to work experimentally and carefully toward improving the human condition. In attending to the dangers and possibilities of genetic enhancement, we will also see genetics in the context of a broad range of options for the community and for parents, not all of which depend on success in a laboratory.

Acknowledgments

It takes a village to write a book. My colleague Arthur Caplan, Director of the Center for Bioethics at the University of Pennsylvania, was more than supportive, his prescient advice was matched only by his willingness to mentor me through a scholarly project for a general audience. John Lachs gave me patient, insightful, and never constraining guidance; his is the idea of a pragmatic approach to applied ethics. Richard Zaner pored through the manuscript and helped me to fundamentally rethink my understanding of the role of philosophers in the clinic. Dick Lewontin has kept me from radically misunderstanding population genetics (or at least tried) and fostered the development of my ideas about genes and the environment.

I was also uncommonly fortunate to have been provided with the resources of the University of Pennsylvania Center for Bioethics, as well as the insight of my colleagues in the Institute for Human Gene Therapy and the Department of Cellular and Molecular Engineering at Penn, and the help of many students and others whom I met only briefly, but who shared a story or insight that proved important in developing the tropes of this book. My colleagues Mildred Cho, Barb Weber, and Charles Bosk influenced my thinking about genetic testing and gene therapy, and colleagues Peter Ubel, Jon Moreno, Pamela Sankar, Paul Lanken, Jon Merz, and Paul Wolpe were very supportive and frequently engaging; Renée Fox insisted that I be rigorous in thinking about the importance of the connection between bioethics and social science. The Leonard Davis Institute helped with insurance research. I also thank Eric Juengst, Peggy Battin, Stuart Finder, Mike Hodges, Micah Hester, Mary Mahowald, Erik Parens, Dan Brock, John Robertson, Adrianne Asch, Herman Saatkamp, Caroline Whitbeck, Mark Fox, Christie Allen, Jeff Tlumak, Charles E. Scott, Lisa Bellantino, and my former colleagues and students at the University of Massachusetts and Belmont University, many of whom helped foster these ideas. Those who patiently helped me wade into molecular work include Ellen Wright Clayton, Arri Eisenstadt, Michele Dewey, Michael Bratman, Jeff Murray, and Jay Horton. My father, theologian Dan McGee, as well as religionists Jean Bethke Elshtain, James Fowler, and James Gustafson contributed a great deal. And many American philosophers guided me through the arduous task of reconstructing James' and Dewey's epistemology, logic, and moral philosophy, including my friends John McDermott, Stuart Rosenbaum, John Stuhr, Kelly Parker, Jackie Kegley, Tony Graybosch, Phillip McReynolds, Pat Shade, and Ken Stikkers.

The National Institutes of Health allowed me, though a summer ELSI fellowship at the University of Iowa Genome Center, to spend a summer working and thinking about the technical part of this work. Harvard Widener Library provided space and support so that I could expand my research on the biotechnological industry, screening, and philosophy of biology. I am also thankful for support from the Hastings Center, Kutztown University, the Conley Humanities in Medicine Seminar at SUNY Health Sciences Brooklyn, Miami University of Ohio, California State University, the University of Tennessee, Temple University, Trinity University, and the Vanderbilt Program in Social and Political Thought.

This book simply would not have come into existance without the help of Carla Keirns. Her insightful attention to my treatment of issues in the history of medicine and science cannot be overstated. My Senior Research Assistant, Dr. Silke Weineck, also worked through numerous drafts. Before we made the decision to withdraw and rewrite the work for a general audience, an anonymous reader at Oxford was very helpful, as were Jennifer Ruark and Julie Kirsch at Rowman & Littlefield.

I can only begin to thank my son Ethan, who created the meaning and possibility of this journey without saying a word. Monica Arruda helped me to rethink family and connectedness. And of course my thanks are with my family, who supported me when things were tough and helped me to understand these ideas of hope and choice.

Philadelphia, Pennsylvania
August 1996

1

THE LANDSCAPE OF GENETIC TECHNOLOGY

Catherine and Clay Johanson don't want to answer the phone. It rings again. He's in the back yard, where he can pretend not to hear anything over his weed trimmer. She attends to their son, Paul. On the other end of the line, Claire Redman, a genetic counselor at the University of Washington in Seattle, can anticipate the conversation. She doesn't need to tell the Johanson's about cystic fibrosis. The difficult hereditary disease has already visited its agonies on the family. Their son, now three, celebrated his second birthday in a CF clinic at Emory University in Atlanta, where pulmonologists, dieticians, pediatricians, and a team of special nurses were working around-the-clock to keep him from drowning in the thick mucus that clogs his lungs. This morning Paul is back at home, and Catherine fills his belly with sweet potatoes and apple juice. She is thirteen weeks' pregnant. Answering the telephone means being told whether or not the baby she is carrying will also have the disease. It is a difficult choice they are about to make. Their friends hesitate to talk about it. Catherine and Clay have spent a couple of tough nights over coffee with Father McBry of their Bellevue, Washington, church. Today, the wait ends.

In Newton, Massachusetts, Rhoda and Michael Salvano are also waiting for a phone call. Michael doesn't have much more time. He is awaited at work by his team of young, ambitious architects, each of whom needs his approval to move forward on some chunk of the 63-story building they are to conceive, build, and follow in São Paulo, Brazil. In the five years they have been married, Rhoda and Michael have made all the normal attempts to start a family. They made love daily for months. They timed their activities, he wore boxer shorts, he took the right vitamins, he drank lots of water. If Aunt Maria said that Tuesdays were good for conceiving children, Mike and Rhoda took the day off. In March they gave up and made initial forays into the world of adoption when neighbors recommended an assisted reproductive technique called in vitro fertilization (IVF). Their reproductive endocrinologist, Dr. Charles Madrigliani, assured them that their problems are common and that assisted reproduction might indeed save the day. However, he told a stone-faced Michael that the motility of his sperm was so low as to present a real threat of failure. IVF is already expensive, he explained, and Mike's insurance would not pay for infertility treatments or assisted reproduction. But in Rhoda and Mike's case, the only real clinical option that would allow Mike's sperm to fertilize one of Rhoda's eggs is Intracytoplasmic Sperm Injection (ICSI). Through this procedure, an infinitesimally small glass needle could be used to "help" one of Mike's weak and struggling sperm into an egg harvested from Rhoda. Before the couple could get excited about this last hope to have genetic children, he pointed out that the whole package, ICSI and IVF, could cost upwards of half Mike's annual salary for *each attempt*, and that more than half of the attempts to use these procedures to implant a fetus fail. Moreover, many couples continue to pursue adoption during the waiting period, which can cost tens of thousands of dollars as well.

Mike and Rhoda decided against ICSI. The cost was too great and the results still uncertain. Mike wondered if this new procedure would result in a child born of sperm that nature had not certified. He dreamed of a son who was lethargic. It hurt. Rhoda tenderly approached the question of how important it really was that the child be their genetic offspring. She was adopted, and had always wanted to have a child—to have someone—who looked a little or acted a little bit like her. But she too wanted to make the right choice. And when one evening she caught a television special, she knew they had found the answer. The answer, she told Mike, was the Herman J Müller Repository for Germinal Choice in California. And so, today, after months of hard work of one kind or another, Mike and Rhoda waited for a conference call with the California sperm bank, which has frozen the sperm of geniuses, athletes, and scientific notables from around the nation. For less than the cost of ICSI, Mike and Rhoda hope to give their child both some genetic inheritance (from Rhoda) and the best possible genetic start.

In Philadelphia, scientists at the University of Pennsylvania plug away. Behind the ivy-covered walls of one of science's oldest fortresses, the

Wistar Institute Building, a quiet battle is being waged against dozens of diseases. These are the white-coated soldiers of the Institute for Human Gene Therapy, home of the most important part of the world's attempt to use genetic interventions to cure diseases. Here, Dr. Jim Wilson and perhaps a hundred other geneticists go through technical procedures that have become almost a mantra. It looks pretty mundane to the casual visitor. But for Wilson, scarcely forty, it is the final frontier of gene therapy. The goal is to develop what geneticists call "vectors."

Vectors, as we will see, are the vehicle that will transmit genetic information into the cells of patients. They are mostly modified adenoviruses—the flu reconstructed to make you well. When you get Wilson to talk about the possibility that they will work, he beams. The right vector could help give Catherine and Clay Johanson's son a chance at a normal life. Paul could use an inhaler to suck in the viral vector, catch the modified flu, and slowly receive genetic information programmed into the vector, which could change the infected portions of his lung so that they no longer secrete the thick mucus that requires his parents to pound his back and that will otherwise, eventually, kill him. But the search for this "right" vector is more than a tough battle. One of the dozens of lab assistants has a button that puts it best: "does anything but shit ever happen?" Present inhalers for CF patients have failed again and again. At normal doses, the modified virus seems to fail in its task of transmitting enough good genetic information to make a difference. To get a high enough dose of the vector, inhalers have to be ramped up so high that they pour a cloud of the flu virus into the already clogged lungs of the patient. This is what occupies Wilson and his colleagues this morning, as they too wait for a telephone call from the Food and Drug Administration about whether it is time to move to clinical trials of the next vector for the next disease.

❧

It may be a new world in Washington, Massachusetts, and Pennsylvania, but thinking about human heredity is nearly as old as thinking itself. And the tough choices of the new genetic era are really not much tougher than the choices faced by parents every day around the world. Everyone must deal with the issues of parenthood, even those with no children and those with foster or surrogate parents. In our schools and in our churches we worry about the meaning and scope of parenthood. In the halls of power and around the dinner table we worry about the personal and political meanings of parenthood. Hillary Clinton reminds Americans that it takes a village to raise a child, and Marian Wright Edelman reminds us that even a village can abuse them.

Nor are we just now learning what families are all about. We develop our ideas of the family from children's books and from *Mr. Roger's Neighborhood,* and as adults come to question the normalization of different kinds of families. Is a good father like my father or not? How much of my own experiences do I want to pass along? We also ponder distinctions between natural and learned identity. Why do we look, and act, like our parents and ancestors—and in what sense are we ever free of our heritage? There are stories to guide us here as well, for our culture is filled with them. And we have more rigid characterizations of the good family. Our courts issue decisions about who counts as a mom or a dad, and what kinds of responsibilities come with those roles. If you don't vaccinate your children, you may be compelled to do so. If you conduct yourself inappropriately with your children, society will take over the role of parent. Businesses sell us ways of improving our families, and churches offer us models of virtuous parenting and even virtuous childhood. Eastern and Western cultures have developed extensive and intricate ways of thinking and talking about heredity, and folkways that acknowledge the importance of biological parenthood to culture.

We have been thinking about the biological dimensions of what has come to be known as "heredity" for a long time. Our efforts to control it systematically began with the cultivation of crops at least ten thousand years ago. Human domestication of animals involved both the elimination of weaker cattle and the use of selective breeding. It was used on sheep, goats, oxen, camels, and other animals on the African continent. Assyrian records indicate that as early as 5000 B.C. the use of artificial fertilization was common in the cultivation of date palms. Gradually, the manipulation of animals and plants became an important part of economic growth. The economy of Troy was based on horse breeding.

Study of *human* heredity has historically been linked to social and medical concerns. Diverse traditions maintained that "blood" is of importance in illness, and in social and familial affairs. The Talmud makes mention of the inheritability of hemophilia. Members of the tribe of Levi exclusively inherited the Jewish priesthood. Hindu castes are based on the assumption that "both desirable and undesirable traits are passed from generation to generation."[1] Several Native American tribes hold that the maintenance of tribal integrity hinged on the restriction of intertribal marriages.

Western theories of human heredity were first recorded in the Greek doctrine which asserted that sperm carries hereditary information and "vital heat" from father to offspring. The sperm thus directs the form of the baby. Aristotle disputed the notion that females had the vital heat necessary to contribute to the form of the offspring, and also held that traits acquired by parents during their lifetime might be passed to offspring. The theory that experiences acquired during life could be passed to offspring helped Greeks account for strange differences in appearance among parents and children. For example, Aristotle postulated that a child whose eye color differed from both parents might have acquired the trait from parental experiences.

Despite all these efforts to predict and control the reproduction of plants, animals, and people, the real explosion in the study of the biological family dates only to the past two hundred years. Right alongside has also advanced the practical power to effect changes in families. We have come to think of ourselves as having control over much of relatedness and the family. We talk about a concept called social and biological *identity* of children, a notion forged through years of advancing disciplinary study of inheritance, relation, and what counts as a good baby.

The explosion in the modern, and eventually molecular, investigation of heredity occured in the early 1800s, when research focused primarily on the problems of inheritance in plants important to a large commercial breeding industry: how do the offspring of a flower keep the structures and appearance of their predecessors? To answer such questions, scientists sought to uncover laws of biology, applicable to all organisms, that would explain both inheritance and development.

The introduction of the microscope in the 1600s had led to the discovery by Robert Hooke that plants and animals were made up of *cells*. Cells seemed to provide a kind of matrix for biological life, comprised of independent nodules of activity bustling within every living organism. It remained impossible, though, until the late 19th and 20th centuries, to explain the role of heredity in cellular functioning. How and why do cells divide, reproducing the information crucial to a stable identity in any organism?

Gregor Mendel's experiments with garden peas, which began in 1856 and were published in 1866, began a discipline called *genetics*, which concerned itself with the relationship between traits in the parent and traits in the offspring. Mendel fertilized hybrid peas and observed differences effected by mixing and matching. He identified certain traits, such as height and color. Mendel noticed that a "recessive" trait would vanish from the second generation of plant offspring, then reappear in the third generation in a ratio of one to four. He postulated that a biological explanation for observable differences in offspring must exist, and called that foundation the "formative element."

In 1869, Johann Friedrich Miescher discovered what he called "nuclein" (which we today think of as DNA) while working on white blood cells. Nuclein could be distilled from the nuclei, a gray precipitate that seemed essential to the nucleus of cells. E. Zacharia and Walter Fleming began to make connections between heredity and this material; Fleming observed that the material is also present in fused sperm and egg cells. We owe the discovery of chromosomes, the structures which were formed of this nuclein, to August Weismann and other workers in this area in the 1880s, but also to the German dye industry whose clothing dyes were used as biological stains (the name chromosome literally means "colored body"). By the 1890s, nuclein, or "chromatin" (finally "chromosomes"), was thought to contain the basic instructions for hereditary traits.

While Mendel, Miescher, Weismann, and those who evaluated and contributed to their work forged ahead in the laboratory, zoological investigation into "formative elements" in heredity was being catalyzed by Charles Darwin. In his *Origin of Species*, Darwin crafted an elaborate explanation of the role that heredity plays in the production of whole organisms. He constructed an account of the relationship between cellular biology and the purposive activity of animals.

Darwin formulated the principle of "natural selection." He observed that most animals "produce more progeny than can reasonably survive."[2] Differences among offspring make them more or less suited to survival in a particular environment. The principle of natural selection dictates that organisms with traits more favorably suited to the environment will reproduce more frequently, and more of the offspring will survive—preserving traits that are conducive to survival in a particular environment.[3,4] Over time, substantial changes may be required for survival. Aggregations of favorable traits may produce a distinctly new kind or *species* of creature.

Natural selection theory was an important step toward the modern account that linked animal and human behaviors to biological heredity. Darwin attempted to bridge the huge gap between theoretical biology on the one side and botany, zoology, and human social theory on the other. But Darwin was "pointing along a route which he could not trace. For example, in the absence of a theory of the gene, Darwin could not explain the maintenance of inherited variation that was essential for [his] theory to work."[5]

Darwin's work encouraged those whose research concerned social and political life. He also catalyzed the application of Mendelian and Miescherian genetics to human heredity. Though it has not often been recognized, one immediate implication of Darwin's interdisciplinary research on genetics was a move by many biologists toward *eugenics*, named at the turn of the century by Darwin's cousin Francis Galton, who argued for "the science of improving human stock" in his work, *Hereditary Genius*. Galton envisioned cultural, societal, and familial planning that would move us toward a "better race of men," produced by a series of "judicious marriages over time."[6]

This systematic, academic discussion of planning better babies through biological science was new, though. Long before Galton brought "eugenics" to the biologists, cultures had been thinking about how to construct reproductive relationships. We have always exercised a measure of control over reproduction in society: it takes two to reproduce, and those two choose each other partly for personal reasons, but largely under the influence of family and community values. We learn what counts as attractive, successful, and desirable within the ethos of a community that has models of the successful family and parenthood. At many times and various places that ethos has been fairly emphatic; long before Galton various cultures were telling families not to have children or whom to have them with.

Early versions of what would later be called eugenics were deployed in three ways, each of which was designed to prevent reproduction: exposure or infanticide, abortion, and sterilization. These three techniques are distinguished from milder means of regulating reproduction in only two ways. First, agreement to terminate or forestall pregnancy is achieved by a political or social instrument, such as a law sanctioning sterilizations or a medical protocol for therapeutic abortion. Second, the techniques involve surgical intervention into the bodies of citizens, rather than acts of verbal coercion.

History reports numerous cases in which societies discussed and employed both clinical and nonclinical techniques. The Spartans left their unwanted offspring to the elements. Plato wrote in the *Republic*: "those of our young men who distinguish themselves . . . [should receive] . . . more liberal permission to associate with the women, in order that . . . the greatest number of children may be the issue of such parents." The Talmud, in recognizing the hereditary nature of hemophilia, required the sterilization of hemophiliacs.

Galton took the early attempts at comprehensive control over reproduction as a model for the application of Darwinian principles to governmental controls on immigration and sterilization. However, the idea of hereditary identity, quantified in terms of "traits," gave Galton a way to talk about these controls in a very sophisticated way, and to link the discussion of good social traits to good breeding. Galton's eugenics was in the service of preserving the "best" traits in the community through intrusive social actions.

It was a successful mix. *Hereditary Genius* persuaded many Americans that eugenic techniques would improve society and halt disease. In his *In the Name of Eugenics*, Daniel Kevles chronicles the rise of eugenics: within a year of the publication of Galton's work, geneticists persuaded Galton to speak to university and public audiences. Galton, then seventy-eight years old, helped the eugenics movement as it began to flourish in Britain and the United States. Talk of a "not-so-eugenic" marriage became commonplace, and social pressure to have a "more eugenic" child emerged. Parents had always discussed good breeding, but Galton gave them a way to predict and control breeding through "science" at a time when the culture was coming to regard biological science as the next great frontier.

Social momentum generated enough interest to form a public interest group for eugenics, which took snapshots and moving pictures of the nation's hereditary profile. A Eugenics Records Office was set up at Cold Spring Harbor in New York. Thousands were strongly encouraged to mail descriptions of their ancestry for filing with the office. Over time, as hereditary information was acquired, family records were to be used for the improvement of communities. If a family incurred too many cases of chronic alcoholism or consumption, they would be counseled to avoid reproduction. Before marrying, a young man would be advised to consult the records.

Whether or not it was effective at breeding better babies, eugenics was more than a harmless fashion. It was changing the scientific landscape as well as the popular perceptions of parenthood and reproduction. The major universities offered courses in "genetics and eugenics," and as Kevles put it, "geneticists warmed easily to their new priestly role." Writing in popular journals, geneticists preached the doctrine of sterilization. Newly legitimate academic and intellectual institutions curried favor among the media and politicians. The public was sold on a new social mission, the "purging" of mental disability and criminality from our hereditary treasury. In order to accomplish the mammoth task, reproduction of the retarded had to be halted. With permission from the legislatures, courts, and public on both coasts, scientists canvassed social institutions that held the criminal and insane. After they were roughly categorized, retarded and criminal Americans were sterilized. By the thousands.

Sterilization received a quick legal affirmation by the Supreme Court: in the landmark *Buck v. Bell*, the Court upheld Virginia's sterilization of a seventeen-year-old third-generation "moron," effectively licensing state sterilization policies. Thousands of institutionalized Americans were sterilized during the 1930s to prevent what Justice Oliver Wendell Holmes termed "generations of imbeciles." With assistance from the Cold Spring Harbor Eugenics Records Office, the American Eugenics Society, and the popularized Eugenics Education Society, laws regulating reproduction and sterilization were passed not only in the coastal states, but in the Midwest and southeastern states as well. Federal immigration laws were also strengthened, and members of certain races (e.g., Asians) were prohibited from moving to the United States.

Eugenicists, eager to find clinical and scientific credibility through universities and hospitals, and power through the government, spoke of a "feebleminded menace of some 300–400,000" in the United States. Eugenicists began graduate training programs in most major universities, and their graduates helped communities—particularly on the east and west coasts—to systematically sterilize the mentally retarded. By 1935, around twenty thousand had been sterilized in the United States. Britain's eugenics program had sterilized perhaps twice that number. These numbers, though, were to be dwarfed by German programs.

❧

The Nazi sterilization projects were the most infamous example of massive social planning concerning heredity. In 1933, modeling his statute on existing American sterilization laws, Adolf Hitler promulgated a Eugenic Sterilization Law designed to sterilize the retarded and their relatives. It was broadened in 1934 to include a variety of "feeble" people:

> Going far beyond American statutes, the German law was compulsory with respect to all people, institutionalized or not, who suffered from allegedly hereditary disabilities, including feeblemindedness, schizophrenia, epilepsy, blindness, severe drug and alcohol addiction, and physical deformities that seriously interfered with locomotion or were grossly offensive.[7]

The counselor of the Reich Interior Ministry "called it an exceptionally important public health initiative . . . 'we go beyond neighborly love; we extend it to future generations.' "[8] Under the Nazi law, physicians reported all "unfit" persons to Hereditary Health Courts, established to determine the sorts of persons who ought not to procreate. Decisions could be appealed to a "supreme" eugenics court, whose decision was final—and could be carried out by force. "Within three years, German authorities had sterilized some two hundred and twenty-five thousand people, almost ten times the number so treated in the previous thirty years in America."[9]

The association with Nazi practices temporarily spoiled Galtonian eugenics in America—notions of a "eugenic marriage" now carried new and dangerous associations with fascism. At the heart of anti-Nazi propaganda was an attack on the eugenic nature of Nazi expansionism, with its emphasis on "purifying" the "Aryan" race. Moral condemnation haunted Nazi physicians, as accusations of eugenic experimentation became a badge of shame. Americans became unwilling to countenance the sterilization of the mentally retarded, and Court decisions gradually narrowed immigration restrictions. Still, the underlying conviction—that complex human behavioral traits could be controlled by social attention to molecular heredity—had been entrenched in biology departments. There, the focus on training human geneticists produced a number of eugenic theories during the 1940s and 1950s.

In the latter half of the twentieth century, new medical technologies were gleaned from advances in the study of biology, molecular and endocrinological. The development of a birth control pill expanded reproductive control, but carried new risks and ways of choosing whether to have children. Amniocentesis, ultrasonography, and chorionic villus sampling (CVS) made it possible to look into the womb to check on a fetus's condition. The possibility of doing so without risk to the fetus (amniocentesis and CVS each carried the risk of inducing a spontaneous miscarriage) through sampling of fetal cells circulating in maternal blood began to loom large. With the 1973 *Roe v. Wade* decision legalizing abortion, diagnosis of a fetal anomaly entailed the new option of therapeutic abortion. These events enlarged the region of reproductive control for families, physicians, and the community. Parents and health care providers were able to participate in social decisions about the traits that are acceptable in a child before that child is born. As the perceptiveness of reproductive diagnosis has improved, the fetus has been more and more open to genetic testing. The presence of a genetic problem (which the media usually misinterpret as the "gene" for a disease) creates a difficult option for parents. Without the possibility of an immediate therapy, and perhaps without the possibility of any therapy at all, parents must decide whether to have a second or third trimester abortion. Society exerts influences on parents as they make their decision, but it is the same sort of influence it exerts on those who are deciding with whom to mate, whom to marry, and when to have children.

In most states, hereditary information is also available to institutions, such as corporations and the insurance industry. Companies whose workers are exposed to chemicals (such as Kodak and DuPont) routinely screen for hereditary sensitivities to a particular chemical in the work environment.[10] Insurance companies have begun to discuss the use of detailed genetic screening of applicants prior to granting health, life, or annuity policies. The fetus of an insured person may not be eligible for insurance, because of preexisting (hereditary) conditions. For example, in 1989 a Portland woman's fetus was tested for cystic fibrosis because the woman's first child had cystic fibrosis. When the test was positive, the insurance company decided that because the child had a preexisting condition (identified before the birth of the child), it would not provide medical coverage for the not-yet-existent who in cases like this are nonetheless already marked as ill. Under pressure from the hospital staff, the insurer eventually backed down. Though no sterilization or immigration policies were involved in this case, it set the tone for an ironic use of *Roe v. Wade*'s description of a fetus as not-yet-citizen.

It is difficult to answer the questions that plague policy makers in the genetic era. Are our genes our property, and no one's business but our own? Or is genetic information no different from any other information used by actuaries? If genes are not our fault, does that mean that other conditions that may be non-genetic or not primarily genetic, such as obesity or high blood pressure, *are* our fault? Will the genetically obese get insurance, while others do not? When do we know enough about the predictive value of a genetic test to move it into the marketplace? Are genetic tests in insurers' interests, or will they rule out so much of illness as to make risk-based insurance an unprofitable business? Will national health care reform ease this pressure or will it only bring about more pressure on parents to use genetic tests that would eliminate offspring with expensive diseases?

Parents are also making difficult decisions about making babies. They make good and bad choices about how and when to have children, with whom, and under what circumstances. The Johansons may elect to have an abortion, in part because of overt social pressures about cystic fibrosis. Financial pressure may convince them that they are unprepared to care for such a child. Where choices exist, parents must make decisions. To leave the phone unanswered, day after day, is still to choose. These decisions draw on their personal wisdom, learned in inculturation and formal education, and on the advice and consent of their friends, leaders, and family. Parents are situated within their community, sharing its values and languages. They apply these highly social values to their decisions, exerting social influence on reproduction.

Regardless of whether the Johansons consciously deliberate about the best outcomes for the human species, the decisions concerning which children will be born (at what time, to whom) are made within the ethos of the community. Thus we can conclude that *parenthood is social.* Parents are influenced by the values of the community in profound ways. Babies are not constructed by rational calculators, they are made by human beings saturated with culture.

The saturation of culture—social and political—is evidenced in the elaborate problem of choosing and consummating a relationship with another human being.

The process of human mating is so complex and diffuse that it is difficult to assign values to the various areas of concern, and almost impossible to adduce the methods of decision making. Thus a "special theory on reproduction" cannot be developed, and a theory of dating and sex that is based on rational choice will assuredly fail. As Jean Bethke Elshtain points out, there is as much theoretical confusion about the creation of children as about any part of human life.[11] Sigmund Freud and B. F. Skinner, Betty Friedan and the Pope, articulate very different accounts of the purposes and parameters of reproduction: it is poetry or politics, biology or free choice, *Eros* or the furtive machinations of baby-making machines. Louise Erdrich writes of this confusion:

> We conceive our children in deepest night, in blazing sun, outdoors, in barns and alleys and minivans. We have no rules, no ceremonies; we don't even need a driver's license. Conception is often something of a by-product of sex, a candle in a one room studio, pure brute chance, a wonder. To make love with the desire for a child between two people is to move the act out of its singularity, to make the need of the moment an eternal wish. But of all passing notions, that of a human being for a child is perhaps the purest in the abstract, and the most complicated in reality. Growing, bearing, mothering or fathering, supporting, and at last letting go of an infant are powerful and mundane creative acts that rapturously suck up whole chunks of life.[12]

But even if our theories turn out to be Rube Goldberg machines, and our advice fails to find any of our friends a date, we still invest tremendous amounts of energy in the discussion and practice of ritualized, sophisticated mating. We subject ourselves to fashions and folkways that can be uncomfortable and awkward. Many of these folkways are transmitted in institutional narratives about proper mating: churches and municipalities authorize the practice of marriage; laws regulate the age at which mating is consensual—and with whom (brother and sister may not marry, nor father and daughter); corporate regulations acknowledge distinctions among kinds of relationships (co-workers must carefully separate romantic and business relationships). Long-standing social habits become laws, practices, and mores, thoroughly "socializing" the decisions that humans make about individual reproductive choice.

Laws and boundaries set by society concerning sex and reproduction are just the most rigid of the billions of reproductive pressures exerted by humans on other humans. More than three dozen teen magazines present stereotypical notions of female flourishing, designed to show girls all of the techniques necessary for membership in one of several "classes" of alluring young women. Read in this way, teen magazines—and much of youth entertainment—circumscribe the boundaries of reproductive success for a variety of cultural groups. Women's magazines promise to help lure the right man. Men have a longer tether, but are still tied to some central fashions and thus encouraged to dress, think, speak, and perform sexually in ways that will be most conducive to finding the "right" person. The drive to reproduce in America is utterly suffused with social values.

From the choice of a hairstyle to the choice of a mate, from the moment she is conceived to the day a daughter moves away from home, trends and pressures create social goals and restrictions. The parent is subjected to the community, and the child subjected to the community and to its parents. Competing values and rules make decisions about reproduction difficult. And these technologies are multiplying at an exponential rate. Every day it seems there is another medium within which we pass along our ideas about what counts as an attractive woman, how to lure one, and what to do once you have found her. The news is peppered with advertisements for vitamins, indoor exercise machines, and plastic hair. The computer is linked to maybe five hundred thousand different sites on the Internet where pictures of men and women are framed in terms of sexual attractiveness. Parenthood is but one part of a growing literature we call "self-help," and so many copies of *What to Expect When You Are Expecting* . . . have been sold that it is possible to imagine the author purchasing an island nation. We have commodified the goals and methods of reproduction and parenting at an alarming rate.

At the same time, new technologies have advanced the study and application of hereditary information for parents. These technologies portend a shift in the use of genetic information. Rhoda and Michael Salvano may not be able to design their descendants just yet, but the power to choose sperm from a drive-through sperm bank is only a few yards short. Where genetic information is now usually employed only to make a simple but agonizing decision about the termination or forestalling of a pregnancy in a family where hereditary disease is present, it may soon enable parents to shop for genetic traits through tests, and enable the parents of those who are ill—or those who are normal—to modify children's genes. In the magazines and books read by most of us, the American parents of a self-help generation, the new genetics seems like an enormous, promising, yet ominous power, a genie out of the bottle.

In 1953, James Watson and Francis Crick published "Molecular Structure of Nucleic Acids: A Structure for Deoxyribose Nucleic Acid" in *Nature*. In that 900-word article, Watson and Crick argued that deoxyribonucleic acid has a structure, the now well-known double helix. This was of enormous importance, because, it was thought, the structure of genetic information would help explain the way that cellular heredity functions and replicates. The story of DNA is the story of a code that our society began to crack. Once you begin to understand and visualize the double helix, your trip into the intricate account of hereditary "identity" is under way.

It is to the helix's twisted structure, Watson and Crick argued, that four kinds of repeating nucleotides attach. These four "bits" of genetic "information"—adenine, guanine, cytosine, and thymine (A, G, C, T)—are decoded into amino acids by the machinery of each cell. Combinations of these bits, termed *base pairs*, make up a single *gene*; a total of 50,000 to 100,000 of these genes comprise the nucleus of each of the body's ten trillion cells. Within the cells, genes are organized in structures called *chromosomes*. Each human cell contains an arrangement of 46 chromosomes (23 from each parent, paired together). Altogether, some three billion nucleotide "bits" comprise the hereditary information, arranged in chromosomes, along the double helix.

Genes contain instructions for the creation of proteins. These *proteins* control the metabolism of cells. Within each cell, particular instructions facilitate the specific sort of metabolism that is appropriate to the function of the cell. Bone cells become bone cells through enzymatic actions, specified by genetic information. As a fetus progresses from one cell to some two trillion cells at birth, the specialized metabolism of cells converts food, fed to the fetus through the umbilical cord, into organ systems and bodily structure.

Though all cells in a given individual have the same genetic information, there is enormous specialization among cells in the body. This is possible because each cell is somehow able to distinguish the particular part of the genetic information that corresponds to its function in the body. For example, a retinal cell is disposed to be sensitive to a certain sort of stimulus because a part of the genetic code enables an appropriate metabolic process within the cell to take place. Groups of human cells, each having the same genetic information, perform different tasks, such as the functions of the liver, or muscle, or hair follicle.

The discovery of the "double helix" helped to explain the way genetic information operates within the metabolic structure of the cell. It facilitated a particular kind of description of the mode of genetic inheritance from organisms to their progeny. But prior to the series of important biological discoveries made during the 1970s, study and control of the information of heredity was limited. Researchers could draw inferences about specific gene expression only from the living or dead whole of the organism. The problem is a foundational one: What is the nature of the relationship between *genotype*, the genetic constitution of an organism, and *phenotype*, the observable physical or biochemical characteristics of an organism? It was not yet possible to discuss the *way* the genetic code, present in each cell, enabled specialization and function. Like Mendel, researchers were limited to the visible evidences of heredity. It was not yet possible to see the way genetic information operated within the cell, or to link that operation to a genetic pattern for the whole organism.

But new procedures were created. Perhaps the most important was a protocol that allowed the *splicing* of genetic information. The splicing process has been likened to a genetic sewing machine. The idea is to take genetic information from one source, mount it atop a delivery mechanism, and insert it into another source. First, DNA is clipped out of the source. Second, a *vector*, which will be used as the delivery mechanism, is constructed of special DNA from a plasmid or virus. The vector has a special mechanism that allows it to insert the source DNA into the cells of the destination organism. Third, the chromosome of the destination organism is then modified by the source DNA. The modified cell begins to follow the instructions of the new DNA, taking as its purpose the duplication of a particular enzyme structure. And, as it duplicates itself, it begins churning out copies of the modified gene.

Using enzymes as a sort of chemical razor blade, biologists were able to isolate and remove specific segments of chromosomal information. When Stanley Cohen and Herbert Boyer successfully implanted chromosomal information from one bacterium into another bacterium, it was an important step toward using the new technologies of genetic information. They had catalyzed the process of genetic engineering, and soon human genetic information would be spliced into bacterial and yeast cells, producing gallons of insulin and other useful compounds.

From a theoretical standpoint, the production of human insulin by yeast cells seemed to validate the connection between genotype and phenotype. If human cellular identity could be reduced to a genetic code, then duplicated by a yeast cell, it would seem that much of what is essential about human biology is located in the genetic information used to create enzymes. It was assumed that what is true for insulin is true for every part of the human body—that there are "codes" for each human function. Genotype precedes phenotype.

As important as these technologies were for the contemporary biomedical sciences, they pointed to a future in which genetic information would be modified for the purpose of a new kind of medicine. Human or animal or even plant DNA might be spliced into the DNA of a human being to cure hereditary diseases or improve human traits. Genetic therapy in human embryos could allow physicians and parents to alter embryos before multiplication of embryonic cells entrenches an undesirable hereditary code. A vision of the future was taking shape, one that depended on a less myopic understanding of human genetic inheritance. Technologies in the 1970s provided a way to splice the genetic information, but gave only scant clues as to the details of the connection between genetic information and animal function. A map of the human hereditary matrix was needed.

By 1984, discussion of the possible technologies for a genetic map of human hereditary information had reached interested ears in the executive branch and Congress. The International Commission on Protection Against Environmental Mutagens and Carcinogens held a conference in Alta, Utah, at which the growing role of DNA technologies in medical research was referred to again and again. A copy of the proceedings made its way to the Office of Technology Assessment, where information from the conference was incorporated into an important research report, *Technologies for Detecting Heritable Mutations in Humans*. This report impressed upon legislators and the Department of Energy (DOE) that a genetic map was essential to cataloguing effects of nuclear radiation on human bodies; further, this map might have medical implications. One scientist even extended the prediction that genetic mapping would change "the future of medical research in the world."[13] Yet the resources and infrastructure necessary for this massive undertaking were not available. Dozens of facilities were needed, each specializing in a small segment of genetic information. Each of these facilities would need funding, and a structure would have to be created to coordinate and oversee the endeavor.

The exact details of such a project began to take shape in Santa Fe, New Mexico, at a massive genetics conference called by the DOE. Virtual unanimity was evidenced in support of a major governmentally funded project to study human genetics. In 1987, the DOE's Health Effects Research Advisory Committee recommended that the DOE "commit to a large, multidisciplinary, scientific, and technological undertaking to map and sequence the human genome."[14] Congress determined that the DOE needed a partner to oversee medical implications, so James Watson[15] was recruited to start a National Center for Human Genome Research at the National Institutes of Health. Together, the DOE and National Institutes of Health were allocated (in FY 1989) $3 billion for the start of a fifteen-year project: the mapping and sequencing of "the human genome."

The Human Genome Project, as it was termed, had seven goals:

- Map and sequence as much of the human genome as possible.
- Prepare a model map of a mouse genome.
- Create data links between scientists.
- Study the ethical, legal, and social implications of the research.
- Train researchers.
- Develop technologies.
- Transfer these technologies to industry and medicine.

To accomplish these goals, twenty university-based Genome Centers were set up around the nation. Each works on a particular piece of the human genome map, and on parts of the six other objectives.

The project is as politically fascinating and theoretically problematic as it is ambitious. At first spearheaded by the peculiar Watson, the project has since seen two tumultuous changes of directorship. Ethical, Legal, and Social Issues (ELSI) research is funded by a subdivision of the project itself, through a 5 percent allocation to bioethics. Art Caplan, director of the Center for Bioethics at the University of Pennsylvania, called this generous allotment the "full employment act for bioethicists."

Theoretical controversy has rained on the Human Genome Project. One central debate involves a core assumption of the massive project: the

Genome Centers hope to construct a unified account of the functional relationship between phenotype and genotype in all human beings (hence "the" human genome) from research on dozens of different individuals' genetic materials. This method assumes something many population geneticists are reluctant to grant, namely that genes do the same things in all populations and environments.[16] If it should turn out that genetic information varies tremendously among populations of organisms, a wholesale map of genetic information from many individuals of varying ages and experiences may not be very comprehensive or accurate. Moreover, if genetic components of diseases differ across ethnicities, ages, genders, or experiences, the usefulness of the entire map could be undercut—human genetic diversity might not allow a constancy across populations. Despite these concerns, molecular geneticists in the project are inspired by their astonishing success in locating a single code for some human enzymatic structures. If areas of genotypic variance are limited to "surface level" traits, such as the color of our skin, they hope that the map will be able to provide clues about many diseases that affect us all, and perhaps the answers to much deeper questions.

The project is divided into three phases: mapping, sequencing, and application. First, the centers will map the approximately one hundred thousand human genes, with each center working on one or two chromosomes at a time. Even as this project will near completion, the second process begins: determining the sequence of each base pair of nucleotides within each gene. There are approximately three billion of these As, Cs, Gs, and Ts, and the task of assigning each one to its respective genes, chromosomes, and functions will be much more vexing than the initial map. Without a Rosetta stone, scientists are forced to draw loose correlations between genetic information (the genotype) and bodily form and function (the phenotype). Why is the gene for Alzheimer's disease also related to digestion? What does the breast cancer gene *do*? This is made even more complex by the vastness of the data to be gathered. Of the three billion-bit encyclopedia of information, scientists *guess* that approximately 10 percent—or less—plays an active role in the creation of bodily structures, or the regulation of cellular activity. The other 90 percent may be "junk," a kind of biological noise that plays no role whatever. Sequencers will need the wisdom to know the difference, wisdom that depends on a comprehensive understanding of the enormous map.

The project has already had a dramatic impact on medicine. Virtually every day we hear of the gene for some disease. As we will see, the presence of these "genes" in the media is frequently not a sign that a test has been developed, or even that a gene has been correlated with a disease in the general population. Genetic tests, once available, make it possible to diagnose someone even when no disease is actually present, and can provide a test for diseases where no cure is available.

Options for therapies and other use of genetic information are also under investigation, or in clinical trials, around the world. When a causal link is found between a gene and a disease, gene therapies may be attempted. Since 1990, several hundred gene therapy protocols have been approved for testing by the National Institutes of Health. These therapies are of at least four varieties. The first involves the use of genetic technologies to provide whatever hormone, enzyme, antigen, or other protein the body fails to provide, but leaves the genetic material of the patient untouched. Genetic engineering of bacteria to make human insulin, which is then injected by the diabetic, is an example of this first kind of genetically *related* therapy.

The second variety of therapy is also directed at curing a condition without modifying the patient's genetic makeup. It involves the use of genetic engineering to create a variety of unusual biological materials that may be used in more or less invasive therapies. The University of Iowa is attempting such a treatment of cancerous brain tumors:

> At the University of Iowa Hospital doctors hold out hope for 15 brain cancer patients about to undergo treatment with a herpes gene. Investigators believe that the gene, once injected, will burrow into cancerous cells and make tumors vulnerable to ganciclovir, a herpes drug. In theory, the tumors should be killed, leaving the surrounding tissue unharmed.[17]

The third variety of genetic therapy is to more properly engineer the patient's own genetic material. Therapies of this third variety modify only the patient's *somatic* cells, cells not involved in reproduction, and thus do not affect future generations. Several current genetic therapies for cystic fibrosis are of this variety, involving the inhalation of modified genetic material suspended in a cold virus. If the patient catches the cold, the virus will, it is hoped, temporarily replicate the healthy DNA throughout the lungs. Another example is the transplantation, at Baylor, of genetically engineered liver cells into children with end-stage renal failure.[18] As we will see, the third variety of gene therapy requires a quantum leap forward in genetic technology.

The fourth class of genetic therapy involves modification of the *germline* or sexual cells of the patient. These modifications of reproductive cells would be passed on to the genetic descendants of the individual. No germ-line therapies are in progress in the United States, and both the Human Genome Project and the world organization of genetic mapping efforts have placed temporary moratoria on such protocols. In fact, during 1992 the Human Genome Project considered requiring that all applicants for genetic therapy protocol review at NIH certify that therapies could not "lead to germ-line engineering." The proposal was abandoned in 1993.

Surrounding these four specific genetic therapies are a host of new and related reproductive technologies. The Salvanos considered ICSI, in which a sperm is directly inserted into an egg, which may make it possible for virtually every man to use his own sperm to reproduce (regardless of sperm motility, which had previously been essential in IVF). Recent developments in veterinary medicine involving the preservation of stem cells suggest the possibility of extending the viability of a man's sperm almost indefinitely. Embryo extraction may make it possible for a fetus to be removed from the mother's uterus to give it up for "frozen adoption." The development of an effective artificial womb for animals (most recently including goats) has been accompanied by the reduction of nearly a month in the age of viability outside the human womb. Neonatal intensive care units (NICUs) have rolled back the age of possible infant survivability by nearly a month. Could a fetus be grown entirely outside a human being? Likely so, and perhaps within the next twenty years.

Women may also soon have the opportunity to freeze not only fertilized embryos (already common), but eggs as well. A fifty-year-old woman could then have access to her own healthy eggs frozen when she was twenty, thereby enabling people to wait to give birth until they are ready. A grown man might be presented with his baby twin as a "gift." The development of these correlative technologies further enhances reproductive choice, and heightens the public sense that "genetics" writ large is changing the world. While the genome project itself is specifically responsible for only a few spin-off technologies to date, a modicum of success in sequencing and mapping human traits creates significant new choices for parents in the world of assisted reproduction and genetic testing.

2

THE MAGIC ANSWER? HOPES FOR GENETIC CURES

The Human Genome Project is enormously expensive, in the league of the Apollo Project. With the genome project, science turns its attention from the outer reaches of the universe to the most precious of our possessions—ourselves and our posterity. The scope of the genome project is as ambitious as that of the Manhattan Project, involving numerous independent researchers, centers, technologies, and strategies. The best technologies and staffs in the world have been brought on board for the project. It is not surprising, then, that the project is hailed as among the most important scientific endeavors in history. Hopes are high for human genetic engineering.

New Genetic Diagnoses and Cures

Dozens of researchers head Genome Centers in major universities, coordinating huge teams of investigators, each of which works on a particular part of one or two chromosomes. These team leaders share a confidence in genetic research not unlike that shared by Francis Galton with his followers in the 1930s. Thus Thomas Lee writes that "the effort underway is unlike anything ever before attempted . . . if successful, it could lead to our ultimate control of human disease, aging, and death."[1] One prospect, "negative" genetic engineering, is the use of genetic technologies to diagnose or cure many diseases and conditions. Leroy Hood, director of the Genome Center at the University of Washington, forecasts a future in which the project will reinvent medicine. Hood writes that genetic therapies, using human chromosomal information to repair tissues, are only the beginning—the Human Genome Project will move medicine from a reactive to a proactive position, preventing disease rather than curing it. Hood is articulating the trademark hope that, as a result of screening and gene therapy, children will no longer be born with disabilities. Congenital diseases would be purged from the gene pool.

Proactive medicine sounds great to everyone. Most of us believe that medicine is too focused on intensive interventions rather than the kind of care that can keep us well over the long haul. Proactiveness is everyone's favorite medical underdog, the ideal that parents imagine when they set out to turn their daughters into *Dr. Quinn, Medicine Woman*, but that always gives way to cardiac surgery or radiology or neonatal intensive care. But proactiveness with genetic disease is not the same as maintaining a balanced diet. It is interventionist medicine extended to diseases you do not yet have but possibly will get. Physicians will, Hood thinks, begin to treat patients prior to diagnosis of the disease.

❧

Take Blanche Carney. This past week, Blanche found out from Myriad Genetics (and her physician) that she has tested positive for a mutation of BRCA1, the "breast cancer gene," and thus learned that she is among a population that has a lifetime risk of breast or ovarian cancer of roughly 87 percent. She does not have breast cancer. She does not have ovarian cancer. Yet now she has a new set of choices that once faced only those with the disease. Should she have her breasts prophylactically removed? Should she have her ovaries removed? Blanche cannot know whether she will be one of the 87 percent of those with the mutation of BRCA1 that actually get breast cancer. Even if she knew that she will get the cancer, it is unclear how much the removal of breasts and ovaries will reduce her risk. So these are especially hard choices. What if Blanche were pregnant with a fetus that had BRCA1? Is breast cancer a good reason for abortion? The proactive medicine we have the ability to do with genetic diagnosis is not the bucolic prophylaxis practiced by Marcus Welby. It is draconian stuff.

More of such "well-care" medicine is on the way for patients like Blanche (if she really is a patient). She might be allowed to take Taxol or another endocrinologically active medication that might reduce her risk of eventually getting breast cancer. But these too can have dangerous tradeoffs. She might increase her risk of other cancers by taking the Taxol. In addition, such treatments are expensive and their insurability is unclear.

❧

Moving to proactive genetic medicine requires a new understanding of diagnosis: *one will "have" a disease when one "has" the genetic information associated with predisposition to the disease, rather than when one has symptoms.* Just as the autopsy, and later the x-ray, allowed physicians to treat patients on the basis of measurements and pictures rather than patients' reports of symptoms, the future, according to Hood, will see physicians treating patients who don't have any disease at all—just the biological precondition of a possibly painful future.

If Hood's prediction of a proactive genetic medicine is accurate, the basic structure of illness will be demonstrably linked to DNA, and will turn out to have little to do with patients' reports of symptoms. In this future, patients might carry the "universal genetic ID card" touted on television commercials and in *Mission: Impossible*, which would contain a text file of their DNA. As new diseases are connected to hereditary information, the patient could present her ID card at any physician's office, and find out in advance if, say, breast cancer is in the cards. The emphasis on faith- and New Age–healing will turn out to be as clinically bankrupt as leeching; the real diagnoses and cures will be in the genes.

❧

For Carla Strumpf, that day is today. She got the call early this morning from a researcher at a Pennsylvania hospital dedicated to work on dietary and metabolic disorders. She had been having problems with digestion and metabolism all of her life, and the genetic tests conducted on her two weeks ago promised to make it easier to treat her conditions, and to make it possible to put a name on her pain. She was eager. But this morning, the physician-researcher revealed that the test of her gene that codes for apolipo protein E revealed that she had duplicate copies of E4. This not only has metabolic implications, it also, the physician blurted out, "means you may get Alzheimer's disease." The multiple interlaced effects of genes on traits mean that more and more of these "cross-tests" will occur, as patients are "accidentally" tested for things about which they may not want to know.

❧

Hood's prediction rests on several assumptions. First, he assumes that most of human disease is statistically predictable through hereditary information. So far, only a few of the genetic diseases have been so discrete. There are more than three hundred different mutations that are associated with cystic fibrosis. But Hood forecasts a time when integrated sequencing will show the sophisticated interrelations of all these discrete bits of hereditary information, and open up hereditary components of other diseases, even viruses. Hood believes, as do most of the boldest advocates of the genome project, that hereditary information is identical for all persons and environments: the genes for diseases in my body will be the same as the disease genes in Rwandans or the French. If diseases turn out to have different markers in different parts of the country or for different ages—or if some diseases spontaneously occur as a result of "non-genetic" factors—we will have to return to the theoretical drawing board.

Aspirations of human geneticists for medical advances extend beyond early diagnosis and preventive medicine. The Human Genome Project promises to make more diagnoses possible for more people and for additional conditions. The all-out search for hereditary markers has also focused on alcoholism, depression, intelligence, criminality, sexual orientation, aging, and Alzheimer's, to name a few. The hope is that genetic diagnosis will clarify the causes of known diseases and separate disease clusters into many distinct diseases.

One example of the latter is Alzheimer's disease. At present, behavioral diagnosis of Alzheimer's is usually based on unreliable observation scales. Genetic technology could replace this catch-all category of geriatric medicine with a list of clearly defined neuromuscular diseases afflicting the aged—each with its own genetic marker and possible therapy. Another example is that of depression. With the number of diagnoses at an all-time high, the etiology and chemistry of depression are quite hazy. As a result, psychiatric treatment of depression relies on secondary seratonin re-uptake inhibitors Prozac, Zoloft, and their ken, whose collective effectiveness is just barely above a sugar pill and whose long-term effect on the personality is virtually untestable. It is hoped that genetic linkage studies will uncover the cause of depression, addiction, attention deficits, and other recognized disorders by locating their biological ground. A patient's life history and environment could then be sublimated to the determinative power of the genes. Similarly, psychiatry has long searched for a genetic basis for alcoholism, and more than one prominent (but later disproved) study has linked alcoholism to a chromosomal anomaly. The foundational assumption guiding such studies is that many, even most, diseases are inherited, or at least that predisposition to them is handed from parent to offspring. Even without clear genetic evidence, virtually all of the self-help literature on alcoholism makes empty reference to biological components of alcoholism.

The optimists hope to cash in that reference, demonstrating that indeed alcoholism comes in part from a gene.

Uncovering the causes of other undesirable conditions is even more ambitious: biologists hope to trace etiologies for conditions such as aging. The medicalization of aging and intelligence deficiencies, to name only two frontiers, expands both the territory of medicine and its impact on our lives. Though a treatment for aging has not yet been suggested, the breakdown of the cellular aging process into various hereditary metabolic parameters might shed light on ways to guard human cells against some of the ravages of time. Once "traits of aging" can be identified as metabolic norms, aberrant metabolisms can be classified as conditions or diseases to guard against. Rather than fighting aging, we might merely fight each aberrant metabolic trend. Thus aging, through the diagnostic power of genetics, could be seen as an avoidable disease. Similarly, many hope that new genotypic information might lead to advances in neurobiology.

Hood is not alone, indeed is joined by a roster of Nobel laureates led by Sir John Eccles, in celebrating a future in which deficiencies in thoughtfulness could be cured rather than pitied. The idea is to expand the territory of medical power to help the public. If we can find genes for intelligence, deviations or shortfalls can be quantified and diagnosed or treated.

The promise of understanding other social traits, such as sexual orientation, aggressiveness, and even hostility through the genes is actively being pursued by research biologists. Dean Hamer's team at the National Cancer Institute has done several studies linking homosexuality in males, but not females, to chromosomal patterns on the X chromosome.[2] British geneticists have labored for more than fifteen years in search of a constellation of gene markers for criminal aggressiveness. Recent studies have uncovered a promising linkage between violent tendencies and certain chromosomal anomalies.

These studies promise to provide diagnoses and, later, tools for therapies for social traits. Medicine, Hood concludes, can play a role larger than its present interventions against disease. The physician will change medicine: instead of fixing the broken body, he or she will improve our quality of life through preventive genetic knowledge. Therapies will target many conditions of life not presently open to medical intervention. Just as medicine shifts from intervention to prevention, perhaps it will also shift from negative to positive genetics—either by enlarging notions of disease to include large sectors of social behavior or by expanding the roles of physicians beyond treatment of diseases.

A Perfect Baby

The popular literature can hardly mention the word "genetic" without including a description of engineered, perfect babies of the future. The description of 1990s-style, consumer-culture perfection in humanity is repeated in virtually every major news journal's coverage of genetic advances: 6 feet tall, weighing in at 185 pounds, without hereditary disease. His brain is engineered to an IQ of 150, with special aptitudes in biomedical science. He has blond hair, blue eyes, archetypal beauty, and poise. Neurotic and addictive tendencies have been engineered out, as has any criminal urge, but in the male model, aggressiveness is retained as part of the "athleticism" package: muscular and quick, he is competitive and can play professional level basketball, football, and hockey. He also has the "sensitivity" package, and enjoys poetry from several cultures and periods.

But much better babies are not only the stuff of *Brave New World* and *Time*. The new portfolio of reproductive choices is at least in part what makes for the attractiveness of genetic engineering in the popular press: parents could participate in the scientific and systematic construction of their perfect baby. How might parents make these decisions? The choice faced by the Salvanos, shopping for a sperm donor, offers a clue. At the Repository for Germinal Choice in California, the sperm of "geniuses" and "athletes" is stored in a special bank, teaming the technologies of in vitro fertilization with the hopes of genetic enhancement. The sperm of a luminary or jock can be placed in a petri dish with your egg; several of the fertilized embryos can then be implanted in your uterus. And you don't even have to ask the donor out on a date or read his latest book.

Similar sperm donor choices have already found their way into more mundane in vitro fertilization (IVF) centers across the country. Parents in some medical centers may request information about sperm donors, including religion, age, ethnicity, and a variety of bodily and medical characteristics. This has placed IVF physicians in the position of choosing whether to dispense entirely nontherapeutic information. The power to make that choice enlarged their social role considerably. Even if an IVF team refuses to release certain donor information, it still can and chooses not to. Thus the power to decide about release of information pits a physician's desire to protect autonomy against the physician's own judgment concerning the privacy, inheritability, and desirability of donor traits. No matter how the physician decides, she will still have made the important decision. There is virtually no law regulating IVF.

Decision making about IVF donor information opens the door to other decisions physicians will soon make concerning the use of reproductive technologies. Patients may request tests to screen for a host of conditions. Hamer's homosexuality study involved the creation of a diagnostic probe that physicians could use to look for these markers. Physicians must decide whether to perform this nontherapeutic diagnostic procedure for patients who request information about their fetus's sexual orientation. Whatever their decision, the ability to perform the procedure gives new powers to physicians, power that either enlarges the territory of medicine by pathologizing homosexuality or gives the physician the new, nonmedical position of being reproductive adviser and technician.

Little or no technical modification will be required for physicians to take on these new powers in the area of reproduction. Physicians would use the same sorts of procedures to diagnose homosexuality or aggressive tendencies that are used to diagnose cystic fibrosis or Alzheimer's. The same informed consent provisions would apply, requiring the physician to explain that these procedures may not effectively diagnose the desired condition, and that there may be social or economic implications. The momentum will be toward including some of these new tests in a one-shot "panel" of genetic tests, alongside the many other diseases for which genes have been identified.

As we will discuss at length later, the only real shift would involve the understanding of what physicians may properly do for, and to, patients and society. If a physician agrees to dispense these technologies, it will have the effect of bringing these technologies into the realm of medical practice—even if we call the traits we test for "conditions" rather than "pathologies." The American community's faith, and long-standing covenant, holds that what physicians do is to cure the sick. If a physician concerns himself or herself with nontherapeutic traits and conditions, while acting as a physician in the medical context, the net effect is that the definition of pathology (and thus of medicine) is expanded to include the new area of concern.

Beyond new diagnostic choices for parents and physicians lies the possibility of genetic enhancements through gene therapy. Images of gene therapy for enhancement are everywhere in our culture, though the reality is that genetic tests for the purpose of enhancement will be just about the only genetic enhancement available in the next twenty-five years. Still, engineering the human species through direct and systematic modification holds such imaginative potential that gene therapy enhancement has been celebrated by innumerable writers, each of whom emphasizes the potential for a "human canvas" on which to draw a better being.

Social theorist Brian Stableford is one such writer. Stableford endorses a "new quest for better human beings." Society is ready, he writes, to progress from our present stage of social engineering, which he (following

Daniel Dennett) terms "second phase Darwinism." The present stage began when we became aware of the role of forces of evolution in human conduct:

> In the second phase, most of our achievements in controlling the evolution of other species were accomplished haphazardly. Our ancestors had only a vague notion of what they were doing as they bred all the special strains of domestic plants and animals. . . . Now that we are beginning to understand how DNA works, we are also acquiring technologies for tinkering with its workings, which will enable us to become genetic engineers, and take control of the living machinery of cells and organisms.[3]

The capacity to make adjustments to evolution will hoist humanity into the "third stage" of human biological power. "Men will become masters of evolution, and will be empowered to control their own evolution as well as that of other species."[4] Our activities during the final stage begin with "external activities," such as the cloning of human organs for transplantation, and "curative activities," such as the removal of heritable pathologies from the genome. After these activities are mastered, human biological engineering can move to the active advancement of evolutionary development.

Stableford makes many specific suggestions for the improvement of human bodies through genetic modification. First, we fix frailty. Because we are so vulnerable to the loss of oxygen during trauma, often causing brain damage and death, he proposes "small extra lungs, with a tiny heart and blood vessels . . . added to the throat, to keep the brain oxygenated when the major blood-system is injured."[5] Using animal DNA, we might splice in a better backbone to protect us from strain and fatigue. Regenerative hands could be added, to protect us from accidents with knives and chain saws. And the skin, so frail around our vital organs, could be made tougher.[6] Cellulose-digesting stomachs would allow us to eat "lower on the food chain," so that we could make more of foods like grasses—perhaps helping to alleviate world hunger problems. And we have too much small intestine, an organ he favors truncating genetically.

The new world of human life will extend beyond repairs of existing defects. Why not engineer the eyes of a fly into humans, so that we might be able to perceive a greater spectrum in lower light? The owl can see in virtual darkness. The flashlight fish lights its own way—we could engineer similar luminous bacterial colonies on our cheeks. The bat and whale "see" with an aural efficiency we have only roughly begun to duplicate with technology—why not splice in sonar? And hearing has been all but perfected in a variety of different animal forms. We might study these in search of augmentary modifications as well.

With "better protection against cold and some kind of physiological protection against caisson sickness [the bends]," we might live underwater.[7] Stableford literally draws a picture of such a proposed person for the reader: the genitals may be withdrawn for protection against the cold, the skin is tough and scaly, feet are webbed, and a second breathing apparatus is added. It looks like one of the aliens we see landing on Earth in our Hollywood science fiction fantasies. "With underwater houses and factories made of non-ferrous materials, our descendants might have little cause to visit the land of their ancestors. Two separate worlds could develop. . . ." Stableford also puts new humans in space, with sealed skin and eyes and new oxygen and food storage capacity. To live in zero gravity, a different skeletal composition would be needed.

Studies of the correlation of heredity and intelligence are central to genetic mapping efforts, and hopes for genetic alteration of intelligence are shared by many of the genome project's most distinguished researchers. Naturally, speculation about how this kind of engineering might take place is subject not only to potential limitations of the project, but also to those of cognitive neuroscience. What the mind actually is, and how it works, is as hotly contested among molecular geneticists as among philosophers of mind and psychiatrists. At least since the famous exchange between Edward O. Wilson and Richard Lewontin in the mid-1970s, molecular geneticists have wondered how much hereditary control exists over intelligence.[8] For many, optimism about the biological etiology of intelligence culminates in the hope for much brighter children able to solve problems with élan and speed. The same drive that has led parents to give their gifted children Prozac and Ritalin, the drive to obtain better scores that lead to Harvard admission, presses the drive for genetic improvement of cognition.

French experimental biologist Jean Rostand is among those who believe that genetics ought to be put to an even more constructive social use, one that improves on Stableford's piecemeal bodily repairs and on the improved brain. He writes of a new *kind* of human, the human engineered to embody as many perfections as possible. Rostand's future "super" human

> has been the dream of philosophers from Nietzsche's *Zarathustra*, to Renan's *Dialogues philosophiques* . . . it is probable that [through genetic engineering] we would get, in a few generations, men of more than average intelligence, and possible that among them would be found men superior to anything we have known.[9]

Although Rostand, as do most of the writers who cite Friedrich Nietzsche as an example of "superman" ideology, mistakenly attributes to Nietzsche some sort of biological plan for better people, the general point is well taken: everywhere in constructive thought since Socrates, the dream of a better kind of person has been central to metaphysics, religion, and social thought.

The faith in our ability to socially engineer human natures leads Rostand to endorse a hereditary division of labor. Naturally it favors the university professor. Super-thinkers could spend their time thinking things through, with huge brains and visionary creativity. The thinkers would not have to fight for their ideas; a more aggressive species of warriors might perform that purpose. In justification of this schema, Rostand points to the fact that our society is already stratified. More pointedly, present-day children frequently inherit the occupations of parents. Rostand's scheme is to enlarge parenthood, so that all of the mothers and fathers collectively make decisions about who will inherit which lifestyle and body style. Rostand argues that the moral leap from societal and parental pressure to genetic specialization is a small one.

One specific technology to extend this social control of human nature is *cloning*. Cloning is instrumental in a variety of plans for the improvement of humanity, dating to the earliest science fiction novels. Present technology allows the virtually unlimited duplication of an embryo. The difference, though, between a clone of a two-cell zygote and the clone of a 10,000,000,000,000-cell adult is quite momentous. It is not currently possible to clone an adult or even a child. Advances in cloning technology, it is hoped, will come closer to what we mean when we say "clones"—the duplicate of an adult human individual, possessing all of the extant features of the original. In the clone of science fiction, the brow lines earned during a long summer of writing about a new issue are preserved, as is the acidic sense of humor developed to cope with graduate school. A real clone requires us to duplicate not only structures of intelligence or character, but the total person. Your clone *is* you, copied, as if by a photocopy machine.

Such a clone seems to be the key to Rostand's tiered society, stratified by function and form. Cloning eviscerates nature's hold on randomness—a condition that presently plays a major role in reproduction. In this sense, cloning holds a whole range of new possibilities for social and parental control. It also involves the development of a kind of technology popularized in the 1960s by the "transporter" technology used in the *Star Trek* television series: a person can be reduced to molecular energy, then reconstructed—"beamed up." Such a technology makes of a whole person a collection of information, which seems to get right at the heart of the hopes many hold for the Human Genome Project,[10] hopes for the perfect baby.

Political and Economic Changes

The use of genetic engineering for curative and enhancement purposes could represent an opportunity to fundamentally change the social structure. For some, genetics seems to offer a great opportunity to remove reproduction from the control of particular members of society.[11] Others hold that genetics should be entirely subject to all market forces. The best example of the latter view is that of philosopher Robert Nozick, who writes that a genetic "supermarket" is the only just distribution mechanism for genetic technologies.[12] Just as the wealthy, healthy, young white male "happens to be" more likely to receive an organ transplant or plastic surgery than the lower middle-class Hispanic, the recipient of genetic modification to correct myopia or enhance sexual function is likely to be wealthy and powerful. But the wealthy are necessary to fund the process: if the wealthy benefit first from genetics, it is only because they create incentives for research by investing in biotechnology firms and by demanding a high quality of biological life. Nozick insists that the application of market force to genetics is morally identical to application of market values elsewhere. The middle class will benefit from a trickle-down of these technologies, as human nature is slowly but surely improved. For Nozick, medicine is just another marketplace: the rules of the market, like the rules of natural selection, encourage vitality among all living members of the species.

In *Dialectic of Sex*, Shulamith Firestone rails against exactly this free-market ethos of genetic trade and against male-dominated families. In the pursuit of "the ultimate revolution," Firestone proposes that women free themselves from the principal chains of womanhood: childbirth. Childbirth has cost women dearly, in physical and emotional ways. An artificial womb, accompanied by absolute maternal autonomy and genetic engineering, could free women from men and from patriarchal society. Women will no longer be dependent, she writes, on the male rituals and institutions surrounding "traditional" reproduction. New reproductive technologies could eliminate these male rituals, supplanting them with a feminist socialism. Women should be empowered to create their own fetus, to make choices that belong to women. Women will be able to erode outmoded ways of behaving in society. Women can finally usurp the control of birth from its institutional links to patriarchy. What better way to free women from gendered oppression than to lay claim to a future in which women can reproduce without the chains of marriage, men, and sex?

"Pragmatism" as Control over the Media

Genetic optimists frequently express concern about social roadblocks that prevent the attainment of genetic progress. Kenneth Ryan explains the problem in, "Ethics and Pragmatism in Scientific Affairs."[13] Writes Ryan, "the moral lesson of *Brave New World* . . . [is really] the evil of totalitarian control over science . . . the social order in *Brave New World* is maintained by denying access . . . to the pursuit of new knowledge." Science, he says, must have as much latitude as it demands. But the press and other watchdogs will not allow science to make the progress that is "essential to democracy."

Science needs to be more *pragmatic*, he argues. This means amplifying the successes of science in the press, to "build public confidence in the candor" of scientists.[14] Pragmatism means being careful not to let public perception get out of hand—when the public cracks down, scientific progress is excessively slowed. Ryan cites the example of the onerous requirement that scientists obtain informed consent from everyone for any kind of clinical research, lamenting that "it took just a few transgressions . . . to incur the present level of oversight."[15] The philosophical foundation of this "pragmatism" is an embrace of science for science's sake, and a rejection of "arrogant ethical imperialism that finds a moral issue, real or imagined, at every turn."[16] Scientific pragmatism is the only solution to the intrusiveness of a reactionary public and an incendiary press: in tough situations, pragmatism will shift the burden of moral proof to "those who wish to stifle new knowledge for metaphysical reasons."[17]

❧

The Human Genome Project has catalyzed hopes among scientists, physicians, and lay people that disease will be more easily diagnosed. In fact, a new kind of diagnosis might allow physicians to fight disease before it emerges. Greater power to treat disease may be obtained as genes are correlated to physiological conditions. Physicians will develop and introduce technologies whose use does not involve the treatment of pathology. These technologies will enlarge the space of medical decision making. Many foresee a large array of new genetic decisions about longer life span, improved bodies, and greater intelligence. Rostand predicts that genetics will facilitate new social structures, in which stratification is genetic as well as political. Nozick suggests a genetic supermarket, in which innovation is funded by the wealthy, who will also benefit directly by new positive technologies.

Designs of a vastly better human nature lurk behind the predictions of Stableford and Rostand. Cloning technologies to free members of a select social class and female reproductive emancipation are two examples of these designs, which could proceed much further and in a variety of directions, including those popularized by the new social idea of a genetically perfect baby.

Many do not share these or similar hopes for genetics.

3

PLAYING GOD? FEARS ABOUT GENETIC ENGINEERING

The ambitious effort to map the human genome elicits criticism from people of widely divergent backgrounds and with dissimilar interests. Is the Human Genome Project the best use of limited social resources? As philosopher of science Philip Kitcher has suggested, we do not yet know who will benefit from its successful completion or how those benefits are to be distributed. Many of the fears concerning the genome project are really more diffuse concerns about "meddling" with the clockworks of heredity. Did God intend that humans reengineer their heredity? Will we lose touch with the natural biological rhythms of the planet? Feminists fear that male-dominated science will use genetic engineering to subjugate women; some cultural critics fear the erosion of time-tested values by a genetic marketplace. Nor are these fears the whisperings of a pusillanimous minority: the most recent polls show that 49 percent of Americans would prefer not to be screened for illnesses, and 58 percent think that "altering human genes is against the will of God."[1] Catalyzed by Jeremy Rifkin and others, a massive political outcry has arisen against genetic research.

Jeremy Rifkin's *Algeny*

Rifkin, the standard-bearer and lead organizer of the anti-genetics lobby, has figured prominently in the crucial public discussions of genetics. As Art Caplan points out, Rifkin has "an amazing sense of theater" and is a tremendous organizer. When a large group of theologians objected to genetic technology, Rifkin was quietly behind the curtain, articulating the pieces of a theological argument with which he himself did not agree. When another enormous group of prominent figures, led by Bella Abzug, published a petition in the *New York Times* opposing the patenting of any genetic materials, Rifkin was the person crafting the language to get a bizarre and patchwork group together. An announcement about potential cloning? Rifkin has called the media and is protesting outside the university where the science has taken place. New hearings on funding for genetic mapping? Rifkin is first on the agenda to discuss the risks of biotechnology. Though he may be among the least respected contributors to any scholarly conversation about biotechnology, you have to hand it to Rifkin: he has certainly mobilized the rhetoric and the zeal necessary to scare the general public about the potential dangers of genetics. Virtually no public discussion about the patenting of the breast cancer gene by a large biotech company took place until Rifkin blitzed the media. Scholars were content to debate among themselves about whether complex legal injunctions could be used to prevent Myriad Genetics from essentially owning a part of the human body, a question that however answerable surely deserved public scrutiny. Meanwhile, that ownership was all but secure—until Rifkin began sifting through the media, religious, and political connections so vital to beginning a public conversation.

Rifkin's objections to genetic technology are theoretically rooted in fears of the human commodification and abuses of nature, a theme that, ironically, also holds sway over all of us in the cordless, rechargeable, suburban, couscous, NPR crowd. Over time, Rifkin argues, human technological living has reduced nature to an economic commodity. As a consequence, we have lost our ability to appreciate the real value of natural cycles. Perhaps the most important natural cycles are the reproductive cycles that genetic engineering could interrupt. Thus Rifkin argues that genetic research represents unwarranted human meddling in a biological order that is delicate and self-sustaining.

Like Brian Stableford, Rifkin divides the advent of new genetic modifications into stages. Before industrialization and urbanization, an idyllic "first stage" of reproduction depended on humans' natural tendencies. Humans depended on the cycles of the sun and earth, and followed the natural rhythms of season, food, and life in choosing appropriate times and ways of reproduction. Much was left to chance, and the human species benefited greatly from following natural patterns. But humans followed the ominous, aggressive parts of their nature, and began to harness fire and build cities.

Though Rifkin largely concurs with Stableford's suggestion that the "second-stage" in our history was an era of enhanced understanding of the powers of reproduction and the importance of evolutionary forces, Rifkin's is a darker reading of technological change. The problem is that our enhanced abilities to control our environments and progeny came at a significant cost to the "natural" world. As humankind began to colonize cities and work during all seasons and times, the way in which we understood natural resources changed. As had others who more carefully reconstructed Native American criticism of colonial industry, Rifkin suggests that the idea of "balance" and "trade" in energy utilization had been lost, or rather transmogrified, by the human tendency to overutilize energy whenever it suited progress and profit. In the bucolic first stage, we spent energy only when it was necessary, and only in the context of a circle in which that use returned something to the world. In the second stage we began to desire a pool of energy to be tapped by entrepreneurs who would make life better, faster, and cheaper. In essence, Rifkin is critiquing the idea of human progress through technology that is articulated in John Stuart Mill's *Utilitarianism*. By agreeing to negotiate on the means necessary to solve problems, rather than adhering to certain fixed natural laws of conduct, we had begun to think differently about moral problem solving.

The transition from an agrarian existence to industrial capitalism was the transformation of our energy base from *solar flow* to *solar stock*. Human existence had been conditioned by the rising and setting sun, but with the discovery and harnessing of stored sun, humankind broke through the barriers. People now had a batch of stored fire at their disposal.[2] The use of "stored fire" disrupted natural biological relationships, which had depended on a natural flow of solar energy. *With the exploitation of solar resources came an exploitation of the rhythms of life.*

Change was most dramatic in England, where workers began to migrate from farms into the cities. Landlords forced farmers off shared farmland to make way for large-scale sheep farming and other capital-intensive farming. Landlords could then sell their wool to London's growing textile industry, which employed some of the displaced and relocated farmers. English farmers "gave their backs" to the competitive search for industrial work. During the English boom of the 1850–60s, as wages increased, some workers were able to find employment and better living conditions. Other countless thousands were unable to succeed and died of starvation or returned home unsatisfied; they were "unfit."

The prophet of the second stage of human reproductive technologies was Charles Darwin, "sitting on the sidelines, watching this historical spectacle unfold for six long decades."[3] Darwin reads his nation's industrial success onto biology and zoology:

> After examining Darwin's many notebooks, journals, and formal publications, any disinterested observer comes to an unmistakable conclusion: Darwin dressed up nature with an English personality, ascribed to nature English motivations and drives, and even provided nature with the English marketplace and form of government.[4]

Rifkin argues that Darwin's work is only a biological pretext for English industrialization. Darwin's work, already based on Thomas Malthus and Adam Smith, conceived biology on the model of British social life and thereby legitimized that life as in line with nature and biology. Second stage reproductive control began when the most ruthless kinds of competition could be seen as appropriate growth and economic change. In this context, biological and zoological sciences found new status and new purposes. Following Darwinism, a new effort to unnaturally improve humans had begun. Now that human history was linked to biological evolution, the control of human events would involve control over biological endowment.

The third stage of human reproductive control moves us into a post-Darwinian ethos, in which the completion of "social" Darwinism is achieved. Again, Rifkin differs from the optimist Stableford only in that his mood is more bleak. Our children, writes Rifkin, will shed the "all-encompassing" Darwinian "order of things" and "will see the world with a different lens, one tinted to soften the glare of a totally engineered living environment."[5]

> When we ask the question, What will replace Darwin's theory of evolution? we need only look at what will replace the industrial era in order to find the answer. The age of biotechnology brings with it a new way of organizing nature. And if past history is in any way a guide, this new organizing model is guaranteed to be sanctified by the construction of a new cosmology that explains the organization of nature. . . . A new generation, the first of the age of biotechnology, will rest easy, believing that what they are doing to their immediate environment is compatible with the way that the whole world has always operated.[6]

That our children will not question that biotechnology is a function of years of shifting consciousness. Technology is ubiquitous. The computer changes and dilutes the natural ordering of our experience. Telephones have computers. Washing machines have computers. Alarm clocks have computers. And this technological atmosphere has permeated the way biologists, in particular, think about being: "Biologists have been completely won over to the idea that all phenomena are reducible to information processing . . . nature [is] the storage and the transmission of information within a system."[7]

The third stage of genetic technologies, for Rifkin, portends a philosophical shift from nature as sacred (but controllable) to nature as information. "Living things are no longer perceived as carrots and peas . . . but as bundles of information." This information is stored in the genetic code.

Once humans realized that this code is natural, it became acceptable to modify the code—it was "no longer a question of sacredness or inviolability."[8] Because information is everywhere about us, he writes, we do not notice how the real changes in biotechnology are occurring:

> The concern over a re-emergence of eugenics is well-founded but misplaced. While professional ethicists watch out the front door for telltale signs of a resurrection of the Nazi nightmare, eugenics doctrine has quietly slipped in the back door and is already stealthily at work re-organizing the ethical priorities of the human household.[9]

This stealth eugenics is dangerous because it equates "doing good with the idea of increased efficiency . . . 'good' is defined as the engineering of life to improve its performance." In turn, this shift from an industrial to biotechnological emphasis on increased efficiency portends a dangerous, pragmatic social ethos:

> In place of the shrill eugenic cries for racial purity, the new commercial eugenics talks in pragmatic terms of increased economic efficiency, better performance standards, and improvement in the quality of life.[10]

Lest we be seduced by geneticists' promises of slow steps toward a better quality of life, Rifkin argues that we will never be able to draw a line between positive and negative genetic engineering. A test to isolate a hereditary defect in the fetus becomes a pressure on mothers—a good mother will feel impelled to apply many presumably beneficial technologies to the fetus. With each additional diagnostic test and genetic therapy, we plant ourselves on a slippery slope to "trading away our humanity. . . ."

> If diabetes, sickle cell anemia, and cancer are to be cured by altering the genetic makeup of an individual, why not proceed to other "disorders": myopia, color blindness, left-handedness? Indeed, what is to preclude a society from deciding that a certain skin color is a disorder?[11]

There is a solution, for Rifkin, that returns to the problem of sun-flow versus sun-storage. "Nature has given up more and more of itself so that we can secure more and more of our future. . . . We need to pursue a different knowledge path, a path whose goal is to foresee how better to participate with rather than to dominate nature."[12] The choice is either to "engineer the life of our planet, creating a second nature in our image, or . . . to participate with the rest of the living kingdom."[13]

Genetic Patriarchy

Among the more fearsome descriptions of an "engineered life of our planet" is that of Robyn Rowland in her *Living Laboratories: Women and Reproductive Technologies*. Rowland decries current molecular biology as a "masculine science" that desires to "control the outcome [of reproduction], developing procedures to create 'perfect,' 'unproblematic' people, through sex determination, or through elimination of genetic illness, or through the 'enhancement' of a healthy normal adult."[14]

That science is patriarchal is an empirical and philosophical claim for Rowland. Beginning with in vitro fertilization, she argues, male scientists have increasingly had access to the womb. This access changes the character of woman's control over sexual affairs, subjecting it to the endemically sterile and intrusive gaze of physicians. Of course, the empirical claim made by Rowland—that the majority of these new observers of reproduction, each with goals and purposes and methods that seem quite intrusive, are male—is undeniable. But the more crucial philosophical move is that these men, however help-oriented, extend the dominance of patriarchy over women's bodies.

She makes this claim by showing that visits to a gynecologist, a part of the long process of infertility treatment, are only one part of the dismal world of dealing with men and their motives. There are powerful images of the endless stream of cutting, observing physicians with bright lights peering into the vagina. All of these men share a longing for the resource only women possess:

> Procreation and birth are a resource which women have and men want. All forms of creativity carry a certain power; in this instance, the resource of another human being is created as well as the subject of love and affection. . . . If what was offered to women was a sharing in the joy and creativity and limited power of procreation and birth, they might view men's desire to enter the reproductive arena differently. But as it expresses itself in a destructive and woman-hating invasion of women and their bodies, it can never be welcomed.[15]

Men are deeply alienated from themselves, and want to have the power of making a new human being and giving birth—a power that only women possess. In the process of attempting to reproduce this power in the laboratory, men have made women "the experimental raw material in the masculine desire to control the creation of life; patriarchy's living laboratories."[16]

Sexual selection is the first form of patriarchal control over reproduction through genetics. Genetic technologies decrease the age at which a fetus's sex may be ascertained. The power to predetermine sex in the fetus is at hand. At more than forty clinics in the United States, physicians sort sperm by their DNA volume and electrical charging.[17] This allows parents, with a high rate of success, to elect male or female offspring.[18]

Technologies pioneered in IVF research can also be used to select by sex among fertilized embryos prior to implantation. Though sex selection in IVF is still controversial, Rowland foresees a world in which sex is one among a menu of dangerous choices made by fathers rather than by nature. Rowland notes that in sperm selection clinics, parents also and overwhelmingly choose male offspring.[19] Thus patriarchy can come to reproduce itself. And, she writes, the use of sex selection is a slippery slope to eugenics: "it is so closely related to genetic research, that manipulation of the genetic pool could follow . . . and set a precedent for eugenics."[20] In sum, genetic technologies will enable men to reproduce themselves technologically, rather than waiting on women and luck for results.

Rowland questions the population control strategies that turn on selection of male children—how does this really serve the purposes of birth control? The real motivation for sex selection is the misogynist tendency in men.

Though Americans report that they would rather not see sex selection become available in routine prenatal care, Rowland points out that if—and when—these technologies do become available, Americans will doubtless use them to select more men. The impact of this coming choice could be as dramatic as that future suggested by John Postgate, in which societies "might treat their women as queen ants . . . or as rewards for the most outstanding (or most determined) males."[21] These women, breeders for men and disfigured within society, will be the poor: "more poor women who need the money may become involved in a breeding system to create the male powerholders of the next generation."[22] For Rowland, sex selection will reinforce and rigidify traditional, objectionable gender types and create new and dangerous kinds of male interference in women's bodies.

The selection of sex, though, is only the beginning of a commercialized, governmental invasion of reproduction. The genetic technologies that provide diagnoses for today's isolable ailments are already owned, patented, and sold by enormous biotech firms. These firms sell genetic information-gathering technologies to corporations, which screen for genetic "susceptibility to harm."[23] The standards of corporations, and government, become the standards imposed on reproduction. What was once a silent process in the womb, according to Rowland, has now become a social—and male—opportunity for oppression.

Definitions of illness that are based on genetic diagnosis, linked to corporate and insurance company standards, might eliminate the rights of women to define "acceptable life" for themselves and their children. If individual women are forced by economic, medical, or social pressures to abort or bear engineered children that do not meet society's standard for normalcy, those women will lose the most basic control over their wombs. Women who choose to have imperiled children often cherish the short lives of those children and the experience of loving those children during their lives, and always cherish the freedom to choose. Rowland tells the stories of several such mothers and children who lived with inherited disease and flourished. Christopher survives cystic fibrosis to win a major book award. Though Alex dies at age eleven, he is able to win a class Judo award before succumbing to CF. "Rather than use technological skills to screen prenatally, abort, and genetically manipulate, we could assist those with disabilities to live a more fulfilled life."[24]

Like Rifkin, Rowland portrays human genetic engineering as part of a larger, dangerous human tendency to want to control nature. But where Rifkin longs for the halcyon days of sun-flow, Rowland seeks a new era in which men cease to play dominant roles in social and parental decision making. Both decry the emphasis on efficiency, technology, and improvement of our condition that comes with genetic engineering.

Hans Jonas's New Ethics

Hans Jonas, among the phenomenologists at the New School for Social Research who first brought discussion of biotechnology to a broader audience, also worried about these emphases, which he called "the ruling pragmatism of our time." In "Biological Engineering—A Preview," Jonas articulated his fear that "ethical questions of a wholly new kind" are posed by genetic engineering.[25] He sought a new ethical understanding because genetic engineering was, for him, a radically new kind of activity. "What is normally understood by engineering," he wrote, is the use of technology for the benefit of some user. The experiments of the past kept a distinction between human beings and the subject of their experiments: we did not experiment on ourselves and our futures.

> The advent of biological engineering signals a radical departure from this clear division, indeed a break of metaphysical importance: Man becomes the direct object as well as the subject of the engineering art.[26]

With biological engineering, Jonas's analog to the "third stage" of Rifkin and Stableford, people have come to engineer themselves. This shift portends a new kind of "gambling," in which the experimenter must irrevocably change the status of future human populations.

It is this shift that signals the moral problem of genetic engineering. Humans "cannot recall persons or scrap populations. . . . What to do with the unavoidable mishaps of genetic interventions, with the failures, the freaks . . . these are the ethical questions to be faced before permitting even the first step in that fateful direction."[27] And Jonas was concerned not only with the living patient, altered by germ-line therapies or choosing elective cosmetic engineering. His principal concern was for the "defenseless" public of the future:

> over what and whom . . . is genetics a power? Plainly, of the living over posterity; more correctly, of present men over future men, who are the defenseless objects of antecedent choices by the planners of today. The obverse of their power is the later servitude of the living to the dead.[28]

What right have we to modify the future? What right do we have to "predetermine future men"? What wisdom have we to exercise it? These are the questions that, for Jonas, required a careful examination of the technologies of genetic engineering.

Because negative eugenics and therapeutic genetics, aimed at healing rather than improvement, amount merely to "an extension of preventative medicine rather than a beginning of biological engineering," Jonas was not opposed in principle to their use. Unlike Rifkin and Rowland, his goal was not to prevent genetic power from getting into the hands of a small oligarchy, of, say, male physicians. While the "ruling pragmatism of our time" bothered him, he was primarily interested to find the essentially new ethical problems posed by genetic engineering. Therapeutic genetics was on this count merely a new kind of therapy.

On this count Jonas was prescient. Most of conventional gene therapy and genetic testing, we can increasingly see, is technologically similar (and morally almost identical) to other existing clinical techniques. Gene therapy bears strong resemblance to transplantation. Technologically, it is virtually identical, with many of the same problems and only a few fundamentally new ones, like the vector. Morally, it poses less significant issues than are already present for the transplant team. The issues that are difficult are difficult in part because we do not have the analog to the transplant "team" in gene therapy, with its interdisciplinary investigation of appropriate donors and recipients. Genetic tests are similar to so many other clinical tests: it has dubious reliability, it is sometimes suggestive of implications for future generations, and it is immediately interesting to health care payers and providers for a variety of reasons, not all of which are related to the patient.

The danger for Jonas was directed at the big questions, the GenEthics. Will therapies seamlessly phase into "positive" genetic interventions, aimed at improving human capacities, character, life span, or human nature itself? As genetic co-factors are isolated and catalogued, the human ability to eliminate the undesirable will grow greatly—humans may be able systematically to remove traits that seem dangerous but are also valuable. We are often fairly confident that we know the "bad" traits, but may miss the value of more subtle "positive" traits in the bargain:

> Who is to judge the excellence of the specimens—of the semen and ovum donors, and by what standards? Let us remember that it is much easier to identify the undesirable than the desirable, the *malum* than the *bonum*. That diabetes, epilepsy, schizophrenia, hemophilia are undesirable, to afflicted and fellow man alike, is noncontroversial. But what is "better"—a cool head or a warm heart, high sensitivity or robustness, a placid or a rebellious temperament, and in what proportion or distribution. . . .[29]

When we determine for future "defenseless" embryos what kinds of characteristics are best avoided, we engage in an unjustified assessment of the value of human traits. Jonas fears a shortsightedness that would substitute immediate interest for the "unplanned and variegated wonders of man." The "only certainty" of this endeavor is "the impoverishment of the genetic stock."[30]

The natural fruition of human planning in genetic technologies, argues Jonas, is found in the options of cloning. Cloning provides the ultimate human repository of genetic choice, a bank of options as limitless as the technology's ability to copy desirable offspring. We can create, he fears, dozens of Einsteins, instead of celebrating the randomness and diversity that brought forth Einstein in the first place. By storing the clones of virtually everyone, we encourage ourselves to thaw clones of those who later prove fashionable. Clones embody the sad possibility that we might choose offspring from today's values, rather than giving the future a chance. Parents who have cloned offspring will too easily be able to discount the value of their offspring in terms of beauty, health, politics, sex, and skills. Do we really need more than one Mozart? Jonas decries the general human tendency to think that if a thing is good, more of that thing is better. Particularly where the pained lives of the famous and talented are concerned, Jonas wonders: ought we to substitute a market in clones for the random chance that one of these children will be born?

More important, what of the rights of the clone? The parents that long for a Mozart—what will they expect of their child? Jonas fears that parents who invest in a clone will appreciate only what they pay for, and will not allow the child to experience life as anything other than what the child was born (by design) to be. The child's experience of the world will forever be limited by the knowledge that he is a clone of someone who already had a full and accomplished life. Long before philosopher of law Joel Feinberg wrote of a child's right to an "open" future, Jonas speculated that

> the existing subject . . . has an existential right to certain subjective terms of his being—and these are in question . . . the simple and unprecedented fact is that the clone knows altogether too much about himself. . . .[31]

Genetic engineering thus opens the door for parents to meddle in the freedom of their children in what are, for Jonas, unprecedented ways. The clone is never free to be anything but a clone. This makes of the clone a set of traits and expectations, unlike the human child born through normal means. We must guard against the tendency of genetics to make of humans and future humans a mere product, substituting design and control for moral luck. "The very being of mankind for its own sake loses its ontological ground."[32]

Playing God

A number of scholars, and many lay people, fear that in addition to losing the vital uncertainty of natural reproduction, genetic interventions will take from us the sanctified status of being God-created. Genetics, it is argued, is radically different from other kinds of human medicine, and is tantamount to "playing God." Paul Ramsey was among the earliest to affirm a theological resistance to genetic research:

> Men ought not to play God before they learn to be men, and after they learn to be men they will not play God . . . taken on the whole, the proposals of the revolutionary biologists, the anatomy of their basic thought-forms, the ultimate context for acting on these proposals provides a propitious place for learning the meaning of "playing God"—in contrast to being men on earth.[33]

This argument rests on several assumptions. First, it rests on the presupposition that God designed humans exactly as God wished them to be. Second, it assumes that God designed a system, namely natural evolution, in which we all participate but over which we are to have no dominion. And third, it implies that God's genetic gift to us is embodied in the *specific* biology that we possess.[34] An opponent of genetic research and therapy might adhere to the first and third premises even without granting that God acted through evolution.

These premises lead to the conclusion that genetic engineering constitutes "playing God." This phrase, used in widely divergent and sometimes confusing ways,[35] connotes interference in a restricted realm. We play God at our peril, not only because we risk the condemnation of God, but also because the biological materials that we tamper with are beyond our understanding, linked to the divine. Moreover, as Ramsey asserts, "a whole new ethics follows from surrogate biology. . . . Human virtue and righteousness are now to be redefined in terms of the biological *summum bonum*." This new ethics of total control over human biological sanctity seems strange coming from the human beings who have yet to learn how to "tend the garden of God's creation" already entrusted to them.[36]

Playing God is most frequently associated with the loosely described activity of "genetic engineering." As we noted in the introduction to this book, the number of euphemisms associated with genetics has multiplied at a rate only slightly less than the number of lists of GenEthics. Genetic engineering, because it is already saddled with the idea that engineers are at work and that the goals are circumscribed in the same way we make bridges or tanks, sounds ominous and suggests dangerous incursions into God's territory once teamed with the appellation "genetic." Would gene therapy or genetic testing, genetic transplantation activity or biotechnology suggest such overtones? Either way, the stakes are high. God has a plan for human biology, one that we have only begun to investigate. What will be lost when we interrupt God's plan for human biology? Biology is a part of the sacred tapestry of history, and there is something unnatural and immoral about meddling with its weave.

The Dangers of Pragmatism: Leon Kass

All of the opponents of genetic engineering make reference to the dangers of theoretically ungrounded or narrowly profit-driven values in reproduction. Left to the marketplace, genetics becomes the instrument of industrialization, patriarchy, or narrowly cynical utilitarianism. Leon Kass is very concerned about the momentum of biotechnology and the biotechnology industry. The discourse concerning genetics, he writes, has been reduced to "a utilitarian calculus: we weigh 'benefits' against 'risks' . . . [but] ignore the fact that the very definitions of a 'benefit' and a 'risk' are themselves based upon judgments about value."[37]

Kass wants to expand the discussion of genetic technologies to include larger questions about values. In particular, he hopes to encourage those with concerns about genetics to see that judgments about genetic engineering often rest on judgments about more fundamental social convictions. Thus genetic engineering, he writes, must be situated in its many social contexts: in medicine, in our view of social goods, and in economics. For example, he writes that the benefits of genetic engineering often depend on larger judgments about "prolongation of life, control of fertility and of population size . . . the reduction of aggressiveness, and the enhancement of memory, intelligence, and pleasure." But, in general, we have not seen many ready to embrace such a larger discussion.

Kass fears that the discussion of genetic engineering will continue to be too pragmatic. "Simple pragmatism: will the technique work effectively and reliably, how much will it cost, will it do detectable bodily harm, and who will complain if we proceed with development?"[38] Pragmatic philosophy, he writes, has taught us to ignore larger social issues or the fundamental ethical ability of society to confront them:

> Perhaps the pragmatists can persuade me that we should abandon the search for principled justification, that if we just trust people's situational decisions or their gut reactions, everything will turn out fine. Maybe they are right. But we should not forget the sage observation of Bertrand Russell: "Pragmatism is like a warm bath that heats up so imperceptibly that you don't know when to scream."[39]

Philosophers have piddled with technical issues in genetics, missing the larger social linkages between reproductive technologies and social values. Meanwhile, the technological and industrial machinery "turns up the temperature" so imperceptibly that philosophy hardly notices as the dominant social mores are increasingly eroded.

For Kass, pragmatism is at the fulcrum of this "dangerous way of talking." Scientists and social critics have ignored larger issues about reproductive technologies because they are too focused on the balance sheet:

> The very pragmatism that makes us sensitive to considerations of economic cost often blinds us to the larger social costs exacted by biomedical advances.[40]

Kass wants to replace the questions of regional economics (How much is lost or gained by a given discovery for the residents of his hometown, Chicago?) with investigation into the potential for responsible and irresponsible uses of genetic technologies by parents and society (Who will control what technologies, and for what purposes? Who has the wisdom to interfere in which ranges of genetic information?). The "pragmatic tendency," writes Kass, is to ignore larger social dangers and remain focused on immediate progress. Risks and benefits, though, are even in the short term based on our understanding of what is valuable.

❧

Rifkin argues that genetic engineering is the final step in a human conquest of nature, completing a process that went into high gear with industrialization and the promulgation of evolutionary theory. Rowland shares Rifkin's claim that genetic technologies complete a transition in human history, but substitutes the male conquest of feminine reproduction for Rifkin's textualizing of nature. For Jonas, the extension of engineering to human biology implies dangerous new values imposed on future generations, whose identity and freedom are imperiled by genetic engineering. Unlike Jonas, though, theologians who object to genetic engineering on the grounds that it is tantamount to "playing God" do not seek a new ethics. Instead, they caution against the coming interference in the restricted territory of hereditary information. Kass fears that radical critique of the Human Genome Project will be eclipsed by a "pragmatism of risks and benefits" that only scratches the surface of the deeper social issues involved in genetic engineering.

Each of these writers criticizes the "pragmatic" advocates of genetic technologies. Rifkin sees in pragmatism a new eugenics, less shrill than the Nazi science, but quietly moving toward the same dubious goals of a better human nature. Rowland vilifies science's improvement orientation. Jonas contrasts his "new ethics" with the "ruling pragmatism of our time." Jonas's version of a *Brave New World* is ushered in by an all-too-pragmatic rush to improve future generations.

4

DEBUNKING THE MYTHS

Though most Americans learn of new genetic interventions from television and the popular press, American universities are not blind to the choices that face parents in a new reproductive era. In fact, we have studied these choices to death. Scholars in departments of biology, law, philosophy, gender studies, medicine, and sociology have dedicated themselves to research concerning the ethical, legal, and social implications of new genetic interventions, and have written an enormous amount in the past twenty years. However, like so much of contemporary academic scholarship, these discussions are published in journals and by presses that reach only other scholars, and frequently only other scholars who work within our own disciplinary compartment.

While those of us working on ethics in genetics have attempted to develop some education programs, most of these are targeted at the high school student. The typical ethics training for a graduate student in genetics consists of an afternoon lecture devoted to one principle: do not cook the data.[1] Student philosophers, who at the turn of the century were expected to shape public conversation about such matters, are today trained primarily in highly sophisticated symbolic logic and the philosophy of mind and language, while issues in ethics and society are remanded to the "applied" curriculum.

The public has thus come to believe that academia is asleep at the switch of biotechnology, and for good reason. Public discussion of genetic interventions, when it occurs, typically takes the form of the "gee-whiz" documentary, interlaced with sound-bites from bedazzled philosophers but bereft of serious criticism. Thus, while a great deal of genetic research rushes forward inside and around the Human Genome Project, the rank and file are basically not participating in discussions concerning the future of reproductive technologies until those technologies manifest themselves in concrete products offered to parents and others, at which point it is much too late to start a discussion of the ethical issues associated, for example, with testing a fetus for an obesity gene. We sound a bit silly arguing that the test should not be marketed after we have devoted millions of dollars and countless hours to the pursuit of the gene.

Why has scholarship concerning genetic technologies failed to produce proactive public hearings, sensible commentary, and nationwide public policy debate? The answer is fairly simple. American bioethics has failed to consider what William James, among the last philosophers to play an active role in American political life, termed the "cash value" of our scholarship. Those in American academia have largely distanced themselves from immersion into the field of public discussion.

Philosophical discussions of ethical issues in genetics, mostly written for scholars only, are unlikely to help parents, not only because our typical jargon-ridden dense prose is intended only for specialists, but also because most of us have paid too little attention to the relevance quotient of our work for actual couples facing difficult choices. We have largely ignored our universities' promises to be of service to the community at large, and those of us who are philosophers are most culpable. American philosophers have developed ways to measure rigor and success, rankings and professional protocols that have little to do with the life of ideas in the community. Philosophical scholarship thus tends to settle for truths that work within what John Dewey called "hermetically sealed systems" of philosophical thinking. Our systems claim to be grounded in religion or "pure reason," and we too often fail to descend from our lofty and elegant ideals into the mundane world of everyday experiences, where we live and make all of our important choices. If Jeremy Rifkin, as we will see, fails to think carefully about the technical issues of genetics, he at least writes to the general public, a necessity not yet apparent to many American philosophers or their departments.

To make matters worse, genetic intervention is an especially complex area. Biological understandings of heredity, as we saw in Chapter 1, are at the intersection of science and our hopes, technical in their details, palpably beneficial in their practical manifestations. "Genetics" is not the name of a single or monolithic practice or social issue. Rather, it involves a series of choices made for practical reasons by parents, scientists, health care professionals, and government agencies. These decision makers also play multiple roles, and make decisions in multiple contexts. A scientist may also be a worshiper; a parent is also an insured person. There are institutional interests in the outcome of each kind of genetic research: both businesses and churches have taken actions to utilize or discourage genetic technologies. Each of these groups and individuals brings to genetic technologies a set of purposes and histories.

What is needed is a pragmatic approach to genetic investigation and interventions. While we will spell out what that means from a methodical standpoint in the next chapter, the first part involves putting our best understandings of science and culture to work in debunking the ridiculous misconceptions about genetics that are in play in the media. Only then can we turn to solve some of the problems that scientists and parents actually face, or will soon face, as players in the genetic revolution. We know that many Americans are concerned that genetic interventions amount to "playing God." But if the gene therapy physician-scientist cannot figure out how to change his or her *actions* in response to the God criticism, he or she is unlikely to heed it. How is the geneticist to *avoid* playing God? Is playing God something we really need to avoid? Ethical principles that are not carefully attuned to the biological and cultural contexts of genetics will not guide our scientific actions in intelligent ways. We need to filter out some of the noise.

Our first task is to take a look at how well the prevailing cultural critiques of genetic technology work—those espoused by Jeremy Rifkin, Robyn Rowland, Hans Jonas, and Paul Ramsey. Every day, between ten and fifteen patients visit each in vitro fertilization program. Would Rifkin encourage the Salvanos to construct a different idea of family success, based on the sun's flow, rather than turning to the new technologies that might provide them with a child? Should women follow Rowland's advice and distance themselves as much as possible from the manipulations of men, unconscious cogs in the machine of patriarchal domination? Can Ramsey guide the patient, family, and health care team away from any protocols that might violate God's intent for biological heredity? The first step toward making progress in the public's understanding of genetic interventions is examining how suited Rifkin, Rowland, Ramsey, and others are to the task of guiding real choices about giving birth in a technological world.

Our second task lies at the other end of the continuum, where we find scholars with boundless faith in molecular biology and its technological fruits. What practical consequences attend these views? Leroy Hood and others would have us believe that most of our character reduces neatly to some constellation of biological determinants. How wise is this analysis, and how likely is its espousal to produce good public policy? Is it possible to create Jean Rostand's genetic superman, and if so, would we want to? Can Shulamith Firestone expect women to seek emancipation through genetic engineering and an artificial womb in the same spirit in which

Martin Luther King, Jr. sought emancipation from racism and its power structures? Just as we must decipher the practical consequences of the critiques of Rifkin and Rowland, we must unpack the practical implications of genetic optimism. Overly fulsome spirits surrounding the Human Genome Project could license uncritical research momentum and provide justification for questionable research protocols.

There's No *There* There: Nature Versus Technology in the Cultural Critique of Jeremy Rifkin

Rifkin claims that genetics is a part of the economic and political movement toward commodifying nature. According to Rifkin, the fall from harmonious and successful life takes place when humans first trade the rhythms of nature for progress. Rifkin claims that as people moved from rural living to urban industry, they learned to attribute the consequent savage economic competition to nature. This view of nature legitimized scientific and technological progress that would, if carried to fruition, forever alter our genes and disrupt nature itself. At the end of the transition from rural to urban living stand biologists, poised over machines that will create artificial children in petri dishes—unless an anti-genetics lobby can stop all genetic research.

Rifkin's history rests on dubious foundations. His is a poor retelling of both the Thoreau story and the Rousseau story, with human beings as innocent creatures easily corrupted by bad institutions if they fail to embrace the peaceful parts of their nature. On Rifkin's account, our resemblance to the army ant or killer ape is downplayed. And his account of human nature is similarly bankrupt. Thoreau went back into town twice a week, where his mother did the laundry: "bad" technology supports the activities of even Rifkin. We cannot account for the movements of history through greedy actions of a few bad men whose goal was to corrupt the natural order. Humans have ceaselessly pursued technology, and have paid particular attention to the cultivation of crops of animals and grains since the earliest recorded history. Progress in controlling nature, moreover, has been subtle and gradual, not a leap from agrarianism to bioindustrial society in the nineteenth century.

Because humans have always used tools and technologies, we must be suspicious of Rifkin's claim that particular sophisticated tools disrupt the natural balance. Attempts to cast humans and nature in an idyllic balance, which is disrupted by the new "bad" technology, assume that technology is anything other than the way in which organisms deal with the environment. Moreover, because human organisms are as much a part of the natural world as dandelions, we have to wonder how it is that their activities become "artificial."

Human technologies are no more "artificial" than the bear's hibernating in a cave or the beaver's damming of a river. We live in natural worlds, and we are natural creatures. Our ways of eating, killing, loving, and even dying involve our control over other organisms and our environment. Sometimes this control is quite sophisticated, as in the case of respirators and jet airplanes. But our technology is not limited to high technologies like camcorders and genetic tests. Technology suffuses our lives. Language, clothing, all our utensils, all our devices, and even some of our cultural habits (such as economic and legal routines) are technologies, which provide ways of living in, and controlling, the natural world.

Technologies, in fact, are merely our solutions to the problems of living in the natural world. We discover, over long periods of time, that particular ways of behaving produce particular consequences. Just as the infant learns to control its world through tools and words, culture learns to manipulate circumstances and conditions through techniques and technologies. Thus both a hammer and *Black's Law Dictionary* are powerful technologies. Humans construct elaborate tools for the purpose of satisfying a need. Sometimes technologies work, in which case we tend to continue to use them. At other times, they fail.

By contrast, for Rifkin technology is a dark, artificial force that pulls us away from the idyllic, agrarian garden. But Rifkin must forget his own experience to miss the constant and crucial role that technologies play in human existence of all kinds—including agrarian existence. Technology is unavoidable, and no more artificial than any other kind of human activity. When technologies fail, it cannot be because they were not natural. Our test for technologies is their usefulness in the world: what happens when we use them?

Thus certain questions must fall away from pragmatic inquiry. We cannot ask for a proper balance between humans and nature, because humans are of nature. We cannot have as our goal the preservation of a natural order, because all of our activities will necessarily preserve or create some such order. A pragmatic response to Rifkin begins with the recognition that the best human sciences show humans to be animals, and a *part* of a natural world. We interact with, and are made of, the stuff of nature.

Our disastrous decisions, which destroy ecosystems and alter environments, thus hardly signal a human attack on nature. Instead, our follies show nature's destructive side, instantiated in us. Many animals hunt themselves out of existence, or dam rivers, or feed on whole colonies of organisms. That we are capable of similar destruction shows us to share a primary trait with other beasts in our world. Yet the human learning capacity is also that of nature. Each of our new strategies enlists the resources and lives of all of our constituent organisms (the billions of bacteria living inside our bodies, for example). As John Dewey notes, "the processes of living are enacted by the environment as truly as by the organism, for they *are* an integration."[2]

When our human processes and technologies fail, we regroup. Technologies are reconstructed, and we try to reconstitute the world in a way that more successfully satisfies a plurality of competing demands. Our attempts—and our technologies—are suffused with values. The hammer is raw materials, fashioned into a tool. As a tool, it expresses our values and goals. Thus the choice of a new technology is not instrumentalist, nor are particular technologies value-free.

In life, we are seldom faced with the choice of eschewing technology. For example, let us posit that a mother elects to purchase a car: her choice is a multifaceted stab at possible solutions to multiple problems, made within a world where competing demands and exigencies can doom simple solutions. The mother cannot choose to skip technology. She must instead choose among technological options: walking, riding, renting, buying. The good technology will be the one that best satisfies her need to travel within the conditions of her life. It makes little sense to suggest, as an alternative, that the good option for this woman will be the option that uses the least technology or best preserves nature's balance. No solution to her problem will allow her to avoid entanglements with a technological world.

Thus we must move away from debates about technology that oppose it to nature. Human technology is an unavoidable part of human existence. Distinctions between technological and nontechnological modes of being are both impossible to sustain and pointless to posit. Intelligent debate must focus on *which* technologies are best suited to competing needs in a complex society and environment.

❧

We can acknowledge the value of Rifkin's underlying fear: human institutions have made poor decisions about our welfare and the environment at large. This concern seems warranted, but dealing with it requires us to abandon Rifkin's approach. After all, human industrial and medical technologies have provided valuable tools for the extension and improvement of life. We learned how to farm crops, and thus how to live through drought and long winters. Life expectancies have shot up in the past one hundred years, as infant mortality rates have plummeted. The community is no longer limited to a three-mile radius: the telephone, computer, television, airplane, and automobile have brought tremendous opportunities for many Americans. Rifkin is correct to point out that these advances have been a mixed bag. Unintelligent allocation of resources and senseless discrimination have served as ideological and practical barriers against the advancement of human beings. And, at times, progress has not been progress for everyone.

An incredibly complex world, with innumerable consumer choices and only a relatively few unavoidable political or ethical obligations, has perhaps caused "a general diminution, in each person's life, of knowledge by direct experience."[3] Moreover, as professional and social compartmentalization increases with every passing year, it becomes increasingly easy to pass off responsibility for social choices to the higher authorities: we can report that we are "just doing our job."[4]

In this climate, professionals in insurance and business have no incentive to confront difficult moral issues or to educate parents about the new risks of genetic disclosure. Parents might stumble half-knowingly into decisions about prenatal testing that will change their lives. We can laud Rifkin's courage: his followers do not merely attack genetic engineering in technical journals about philosophy, they march outside labs before television cameras. They attempt to change textbooks. They go to political hearings. They get involved in the community and succeed in starting discussions in churches, civic groups, and schools across the country. If we work toward understanding the complexities of genetic engineering, viewing these technologies as part of the real choices parents will soon have to make, we too can move toward a more engaged polity. Genetic technologies need not be a barrier to this remediation of political experience; in fact, they could help to spark it.

A Biological Garden of Eden? Hans Jonas and Paul Ramsey

When Jonas and Ramsey caution us that we ought not to tamper indiscriminately with the future, or "play God," they have in mind an order of things quite similar to Rifkin's natural balance. They thus face similar problems. Jonas assumes that we *could* avoid "tampering" with the future—and that there is a set of choices appropriate to such a posture. For Jonas, the line between tampering with the freedom of future generations and appropriate parenthood is drawn at the point of genetic engineering. With genetic interventions, he suggests, humans begin experimenting on their future. Ramsey similarly argues that genetic engineering is uniquely powerful, and thus will interfere with God's genetic design.

But what are the other options? Can we avoid experimenting with the future? Dewey correctly notes that "all action is an invasion of the future."[5] Our intricate ways of parenting create the atmosphere within which our children dwell: parents make decisions about their children all the time, from before conception until the death of child or parent. At the technological end of the spectrum, intrauterine surgery, in which the womb is opened and surgery is performed on the fetus, allows parents to correct fetal anomalies. Abortion says to the future, "I will not have that child," and thus makes for a certain kind of future, one which does not include the birth of that child. In vitro fertilization makes a "test tube baby." Parents who smoke while pregnant affect their children. Parents who enroll their children in school make a difference to the future. And parents who have expectations for their offspring exercise profound power over them. Even before we have sex, we choose mates, lifestyles, careers, and commitments. All of these project us into the future and create conditions, circumstances, opportunities, limitations, and values with which our children must live.[6] In what sense, then, is genetic technology a harbinger of a new future? In what sense are these new choices? The answer seems clear: we do not need *new* wisdom—we need the historical and social context of *old* wisdom in order to make *new choices*.

To believe that genetic engineering is more than a technological extension of current decision making, we would need to be convinced that the power of genetic engineering is *ontologically* profound. In other words, genetic modifications would have to constitute a special instance of parental control, subject to different rules and responsibilities. In Ramsey's view, modified genes have the power to undercut God's special biological inscription of identity in each child, and thus genetic modifications threaten God's design. Yet genetic engineering is not the only way that we meddle with our molecular code. By smoking, we increase the chance that our cells will be converted at the molecular level into cancerous cells through a modification in genetic material. Formerly good pulmonary cells develop "oncogenes," which code for the formation and multiplication of cancerous lung tissue. Because of its glamour and technological sophistication, Ramsey and others have tended to separate genetic "engineering" from other decisions that affect genes—for example, the decision to drink south

Florida's brown, mutagen-laced drinking water during pregnancy. By selectively emphasizing the unusual features of genetic engineering, Ramsey misses the ubiquity of analogous decision making in present society.

In Jonas's analysis, the power of genetic engineering is the power to rob future generations of freedom and self-identity. But how is this different from the human who does not consider the health of a potential mate—or the parents who know that they both carry the cystic fibrosis gene but choose to reproduce anyway? In both of the latter cases, parents are projecting themselves and their values onto their progeny. The CF carriers are forced to play Russian roulette, given the statistical likelihood that they will bear a CF child, on the basis of judgments about life with CF. The parent who considers the health of a mate before reproducing makes a similar judgment. We all make choices about the ways we conduct our reproductive lives, and these choices deeply affect the future for our children. Choices after a child's birth are similarly restrictive: parents who elect to send Johnny to piano instruction make choices that continue to infringe on the freedom of children.

Jonas selectively emphasizes the systematic control of genetic decision making supposedly made possible by genetic engineering and stresses its apparent difference from the normal decisions of parenting. But we must be careful not to overstate the component of randomness in everyday romantic and parental relationships. This is a pragmatic point. There is love at first sight, it could hardly be denied. But we teach our children that the immediacy of first love should be followed by a set of decisions that each potential mate can make: we try to think in advance about with whom and under what circumstances we will reproduce. We tell our children to be careful, to wait for the right person and the right time. We do not hesitate to teach the skill of planning for the future. Everyday parental activities present the opportunities for systematic—or at least thoughtful—choice.

The pragmatic question, then, is the difference between ordinary parental decisions and changing the germ plasm of the species. The latter appears more momentous. But we have seen that our decisions about changing the germ plasm are not all high-technology decisions; in fact, choosing a mate accomplishes the task of "humans engineering themselves." The difference is in the amount of control that is possible. Jonas favors the use of genetic modifications for curing illnesses, but fears that such engineering could slip into positive engineering and even cloning. However, genetics is not the moral frontier that Jonas takes it to be. Our decisions about the future of children and the possibilities of biological engineering will necessarily favor improving life as well as curing illness. We have to attend to these social purposes, in their contexts, rather than wait for new wisdom.

The Abuses of Gender: Shulamith Firestone and Robyn Rowland

When Firestone argues for woman's emancipation from man and patriarchal domination by way of artificial womb and genetic engineering, she selectively attacks the values that women attach to the process of pregnancy. In Firestone's utopian biotechnological world, childhood "will be abolished," along with the "artificial" intimacy between woman and child.[7] In its place, a new society of freedom and independence for all would be established. Many women, and feminist theorists, find this argumentation imperceptive. Women's experience of childbirth has been celebrated in profound ways. While some women may choose to eschew the process of pregnancy, the argument that intimacy between mother and child is of no value, or sexist, rings hollow.

Nonetheless, Firestone's charge that pregnancy is radically debilitating highlights the fact that many women in America do not receive paid maternity leave and many more become objects of harassment and discrimination. Without endorsing Firestone's ontological commitment to the emancipating power of genetics, we can learn from her forceful claim that women have been victimized by ignorance and misogyny in the workplace.

Reaching the opposite conclusion about the use of genetic technologies, Rowland makes the same appeal to a feminine force under siege by men.[8] For Rowland, men have attacked nature, and women must rescue it from the calculating grip of male scientists. Human reproductive genetic technologies, she writes, are a part of the "control myth," within which we fight against the natural ends of our lives. At the center of the control are men, who work against nature and natural laws by trying to prevent the birth of diseased infants and creating cures for diseases in vitro.

The problem with Rowland's appeal to nature is that it makes unjustifiable distinctions between control and nature. If it is natural to die, it is surely also natural to attend to (or control) the health of our children and to attempt to prevent their needless suffering. It constitutes no denial of the naturalness of suffering and death to practice medicine. It would be *less* natural, given the behavior of humans and animals, for parents to throw caution to the wind when dealing with reproductive health.

We have also to ask why genetic medicine is seen to be so especially patriarchal. Men are no more involved in genetic engineering than in cardiology, and the outcomes in both fields involve prevention of death. Rowland relies on bald assertions about what is natural, namely the control of medicine by women-centered physicians, and does not discuss the reasons why maleness is supposed to be uniquely dangerous or less ontologically natural. In the absence of such evidence, we can only affirm Rowland's commitment to the value of the lives of individual persons with handicaps. She is certainly right to suggest that people with cystic fibrosis can have valuable and rewarding lives. But it is no insult to these people to suggest that serious genetic defects should be prevented—unless we equate the patient with the disease. Many patients would not resist the idea that the world would be a better place if they did not have to suffer and die from the disease. We can over-romanticize the courage of sufferers.

Scholarship concerning the Human Genome Project has tended, either in condemnation or in endorsement of genetic engineering, to rely on outmoded and poorly thought through notions of "nature," "technology," and "freedom." Such categories do little to help us relate better to genetic technologies. The antidote to such categories is a reconstruction of the discussion that takes note of confluences between genetic engineering and other scientific, parental, and social practices. One barrier to this reconstruction, however, is difficult to overcome. It is what Richard Lewontin has termed the "ideology of genetic determinism."[9] And it is everywhere in our cultural literature about biology.

Eat More Guacamole—Classical Genetic Determinism and the Idols of the Body

Genetic determinism is the view which holds that everything, or at least much that is important about human identity, is determined at the moment of conception and encoded in DNA. This belief is suggested by the history of biology we saw in Chapter 1: humans have long utilized heredity to facilitate control over agriculture and human reproduction. Even with the Greeks, it was held that a child growing up outside a human community would naturally start speaking Greek. However, the *theory* that biology is totally constitutive of social life emerged only after Charles Darwin. Then, twenty years after the publication of *The Origin of Species*, our new biological theory found voice in nineteenth-century literature as the belief that "blood will out."

In nineteenth- and early twentieth-century biology, the promulgation of eugenic studies depended on Francis Galton's initial statement of genetic determinism, which assigned genetic determination to virtually all human traits. The cataloguing of family histories and public health recommendations for eugenic marriage depended on the widespread conviction that biological stock is prefigurative. Molecular genetics, particularly the effort to map the human genome, has frequently applied this thinking to its search for a single genome map. It is hoped that the Human Genome Project will provide stable hereditary data for an ambitious array of traits, including not only diseases but also intelligence, sexual orientation, criminality, and aging. Many hopes for genetic progress depend on genetic determinism, or on views closely akin to it.

Genetic determinism is grounded in one or more of three basic convictions.[10] The first is that *genotype*, the genetic pattern of an individual, determines *phenotype*, the traits that an individual evidences. If this were true, we should expect that every aspect of development would be spelled out as an advance blueprint, which issues in a specific, predetermined sort of baby. But phenotypes vary dramatically in organisms with similar genotype (such as identical twins) that have been exposed to different conditions. To be sure, some traits are directly tied to genotype, including blood type and some diseases. However, "such completely genetically determined characters are the exception rather than the rule."[11] In addition, the phenotype and genotype change over time. The real problem with this first conviction, however, is its commitment to the idea that "genes" and "environment" are distinct entities acting in a compartmentalized way to determine the organism. As we will see, the relationship between biology and experience is actually quite fluid and interpenetrative, so that it makes little sense to suggest that only one determines the other.

The second founding conviction of genetic determinism is that "genes determine capacity . . . [on] the 'empty bucket' metaphor."[12] On this account, genetic inheritance sets boundary conditions for capacities in each person, which are permanent and unchangeable. A person has a certain genetic capacity for a given trait, which she or others may fill to a lesser or greater degree. Jack's intelligence bucket may be filled by the facts he learns in school, but his fixed learning limit is determined by the size of the bucket.

The problem with this metaphor is that it not only takes for granted complex genetic powers over social traits, but also assumes that each human's genetic capacity bucket is fixed and unchanging. Thus it is a doubly problematic metaphor. First, all but the most elemental of human capacities seem, as we will see in Chapter 7, to depend on social purposes and practices. The measure and kinds of skills known as "intelligence" change as social beliefs, values, and needs change, so Jack's particular kind of thinking may go in and out of vogue—and may begin and cease to be known as intelligence. A second problem is that Jack is constantly changing, as the places he visits and foods he eats change his genotype. Just as his muscle tissue changes when he exercises and eats particular sorts of food, so the genetic material throughout his body changes in its composition and activity as he grows, ages, and changes. Thus his bucket, if he has one, would never be of fixed size or shape. Jack's body is as changing as the definition of intelligence.

The third justification for genetic determinism is the theory that "genes determine tendencies."[13] When it is suggested that a person has a genetic predisposition to be overweight or a tendency toward alcoholism, this view is at work. As Lewontin points out, this view seems most effectively to buttress genetic determinism of traits, because it appears to take more serious account of the relationship between genes and environment. This view receives apparent support from the statistical exposition of genotypic/phenotypic relation for some particular trait, for example, 60 percent with a particular gene become homosexuals. This statistic seems to offer support to the "new" idea that there is a 60 percent chance that a particular person with the Xq28 gene will become a homosexual. However, from a biological point of view, tendency theory is only a new wineskin for the old wine of "capacities." In fact, an even more problematic notion of capacity is assumed. According to the tendencies metaphor, we *choose* traits, but are *influenced* by genetic suggestions. Thus we think that we choose, while we are really determined, pushed, or persuaded by the "genetic call of our innards." It is as if the overeater hears a voice from the pantry, which is really projected from within the genome: "eat more guacamole."

At the molecular level, the relationship between a particular gene and the phenotype of an organism is a matter of the way that a particular gene reacts to a particular environment. Because genes have only this simple relationship, it is difficult to suppose that they impel organisms to "want" to behave in complex ways. A gene that produced a "tendency" to alcoholism would have to accomplish an amazing array of molecular tasks, resulting in a feeling of needing to drink. Moreover, evidence that a gene accomplishes this feat is taken from the fact that a gene is correlated with a given trait in a percentage of the population. It is to be inferred that the gene did not tend strongly, or that willpower was too great, in the instances where gene and trait are not correlated. This inference simply will not hold. Statistics about correlation are only valuable at the level of *populations of organisms*. Statistical evidence is not applicable to each individual—either a gene will do X in a particular organism, or it will not. The fact that gene G results in trait T in 60 percent of organisms *does not mean* that there is a 60 percent "tendency" toward trait T in any given organism. To say that there is a tendency is just to summarize a statistical finding—it has no further meaning. Similarly, the introduction of a possibility of fighting against some tendency merely confuses matters. Either genes are determinative of a trait under a particular set of environmental circumstances, or they are not.

The idea of "willpower" that is suggested by "genetic tendencies" simply does not apply. In the case of obesity, we can say that genes influence metabolism, but it will make little sense to say that they give us the desire to overeat yet leave the "choice" to the eater. It is a misnomer to speak of a gene that "disposes" to gaining weight. A gene can only dispose toward a particular aspect of metabolism. What you eat, or even what you want to eat, will probably turn out to be less a function of genes than of the intersection of culture and family.

Who Needs a Ghost in the Machine?

The Human Genome Project has only made faith in genetic determinism more entrenched and more complicated. As we noted in Chapter 1, the genome project is premised on the notion that genetic information remains functionally identical across human populations and through time. Yet "every human genome differs from every other. . . . The final catalogue of 'the' human DNA sequence will be a mosaic of some hypothetical average person corresponding to no one."[14] Not surprisingly, while a few diseases have been identified with stable DNA markers, even these do not work for all persons—and in the case of most hereditary diseases there are as many as five hundred different mutations for each gene. The search for a stable foundation for all of human biology, let alone social life, may be radically retarded by the reality of genetic diversity among and within human persons.

The philosophical and medical implications of genetic determinism are profound because our beliefs about the power of genetics determine the way that we approach complex diseases and traits. Take, for example, the case of intelligence. Many molecular biologists, particularly those working on the genome project, argue that intelligence is biologically determined. From their perspective, what matters about intelligence is not education, social conditions, or social purposes. Intelligence is reduced to a common denominator, such as the speed of neural calculation or a gene that codes for a particular chemical in the brain. Philosophers, biologists, neuroscientists, and social institutions have invested years and millions of dollars in this pursuit of a simple substance or process that is the biological foundation of thought. The stakes are high: success would mean the ability to apply simple gene splicing techniques to human intelligence.

The implications of genetic determinism also make it easy to see why so much fear and hope surround the genome project. Claims of DNA's ultimate power over human life make it easier to revere genes as the "code of life," or to condemn genetics as meddling with God and nature. However, if we do not accept the claims of genetic determinism, our hopes and fears are also considerably reduced. Yet society has accepted strong claims about genetic power with little resistance.

Lay people today find biological determinism easy to accept because, as a culture, we have thoroughly assimilated the idea that humans—and nature—are machines.[15] That we think of ourselves, and nature, in this way is not surprising. Human beings have always written about society with reference to the current level and nature of technology.[16] Factory economies in nineteenth-century England inspired a variety of intellectuals to emphasize the role of competition in nature. This competition became part of an inadequate but pervasive description of the natural order. Thus, as Rifkin points out, when Darwin employed the metaphors of industrial technology in *The Origin of Species*, they became sedimented in language quickly because they were helpful ways of getting a handle on the way the biological machinery of animals and humans works. In other words, Darwin's language helped us understand nature, because he made nature look like society. The natural order, we have come to accept, operates more or less thoughtlessly, eliminating the weak and preserving traits of strength, power, and fitness.

This more mechanical genetic determinism has been fueled by the successes of the Human Genome Project in identifying genes for diseases and conditions. If complex diseases are coded into genes, why not intelligence and other behaviors? Similarly, the successes of agricultural genetics in splicing human and plant DNA into animals are seen by some as evidence that what is distinctly human can be encoded, modified, and transported from species to species. DNA and particularly the genes have become idols and with the Human Genome Project have become key to our understanding of what and who we are. Said to contain virtually all that is important about us, they legitimize a view of human beings as human machines. And the danger is that the more we accept this view, the more likely we are to accept the ethics associated with the economic and political grounding of its conception of nature.

We opened this chapter with the claim that contemporary academic scholarship on ethics and genetics has been ineffective in raising social awareness of the philosophical and ethical issues in genetic choices. In part, this is a function of our plodding recalcitrance about becoming actively engaged with everyday society. The public is very interested, though, in genetic technology and has glommed onto the discussions of optimists and pessimists, whom we have seen neither to understand genetics nor to offer any real alternatives for parents and researchers.

Because of its complexity, situatedness, and urgency, human genetic engineering demands a contextually derived understanding of values. Following the biological, cultural, and complex everyday components of parental choices, we have shown that genetic determinism poses the most pervasive threat to such inquiry. Excessive faith in the power of genetics plagues optimists and pessimists alike, but such faith will not withstand the realities of biological diversity and the biological-cultural continuum. We need a better method of understanding genetics.

We will see in Chapter 5 that while biology does play a prefigurative role in sustaining human genetic diversity, genetic factors are not determinative of human action. Beginning with this problem of genetic determinism, we will examine the supposed relationship between genotype and phenotype, and nature and nurture, which receive so much attention in the literature and which are supposed to license the conclusion that DNA is discrete and determinative. In its place, we will reconstruct the relationship between genetics and the social world, between biology and culture. Biology has a role in human social life, and purposes invariably enter into the descriptive claims of biology. Pragmatic genetics will neither reduce biology to politics, nor reduce political purposes to the drives of biology. A better biology will see the limitations of genetic idolatry as well as the importance of genetic diversity.

Our pragmatic approach to genetics will also need to find continuities between the languages of biology, philosophy, religion, and parenthood, if we are to do a better job of connecting genetics to public discussion than has so far been accomplished. The theory of value that we embrace must make sense in the specific situations in which it is to be employed. The tools of philosophical inquiry are important, and philosophical principles are a part of the context of intelligent inquiry. But ethics is involved with practical choices that make life better, not abstract principles unconnected to daily concerns. An ethics of practical life will also have the benefit of being useful to biologists, parents, and social institutions, rather than to philosophers alone.

5

BIOLOGY, CULTURE, AND METHODICAL SOCIAL CHANGE: A PRAGMATIC APPROACH TO GENETICS

Misunderstandings about human nature, many of which we have seen to be linked to genetic determinism, have caused mass confusion about what genes can and should do. We need a better way to understand the role of biology in our lives.

Fortunately, classical American pragmatism, in the writings of John Dewey, William James, and Charles S. Peirce, was explicitly attuned to the implications of the natural world for everyday life. The idea is that to understand difficult social problems, you must examine both their biological complexity and their cultural roots. More importantly, the pragmatist tradition teaches us that we have to remember that ethical problems are problems for *real* people living today, not test cases for some larger universal theory. Thus the pragmatist works to make sure that bioethics stays linked to the public and that the education of the public is a primary part of the problem-solving process. In the final analysis, solutions to public problems are ethical when they *work*, by helping society to think about, and then accomplish, goals that make both biological and cultural sense.

Many philosophers trained in other traditions literally cannot recognize pragmatism as a *moral* theory. Those in the tradition of Immanuel Kant, for example, come to think of the brute inclinations that come with our bodies as nothing more than instincts that rational humans must fight in order to exercise their real, rational moral ideas. What makes us human, on the Kantian account, is our ability to separate ourselves from the natural world and its inclinations; to use strict, universal rules to fight against the demons inside us. Within such a tradition, pragmatism seems to be a facile adoration of our animal natures or, worse, a simplistic account of biology not suitable for either a philosopher or a geneticist.

But such a position, we have seen, is specious. Those who would condemn or endorse testing and modification of the genome need to understand the basic biological facts, and moreover need to be informed by a whole-cloth theory of social method and of human nature. The failings of contemporary scholarship, we saw in Chapter 4, stem primarily from misunderstandings of genes, cultural roles, and the history of social progress. Our moral theory must be informed by scientific advances, so that our arguments make sense. The idea that we should not play God is wrong not because it has some great, universal logical error but because it is based on virtual ignorance about how much we already play God in medicine and science, and because it is not based on a constructive and scientific account of how to *avoid* playing God. We can do better. What we need are better tools.

In this chapter, we reconstruct the role of biology and culture so that we can make some sense of what we are to do with new genetic discoveries. After looking at biology and culture, we will explain the ways in which biology and culture interact in the bodies and lives of parents and children. Only after we have established this pragmatic method can we turn, in Chapters 6 and 7, to making some personal and policy recommendations about the appropriate uses of genetic technologies for curing illness and improving the human condition.

Biology

Heredity plays a complex role in the formation of human nature. We inhabit the *bios*, the world of living and dying creatures. Our lives take place within various large and small organic environments, and each of us is home to an ecosystem consisting of millions of organisms of varying sizes and sorts. As individual persons, we share much with even the tiniest of these creatures. Like them, we must take in nourishment in order to live. We all eliminate waste, age, and die. Moreover, we are linked to all of these creatures by a web of interactions. We share the food and we eat each other.

As much as we share with other creatures, we are also different. Ants are not human, and humans are not ants, because biological limitations will not accommodate certain kinds of development. It is not possible for ants to develop silicon computers because they are biologically incapable of lifting the required materials. The differences between humans and ants are illustrative of the differences among ways things live in the environment. The environment is in relationship with creatures, providing possibilities:

> The environment of an animal that is locomotive differs from that of a sessile plant; that of a jellyfish differs from that of a trout . . . the difference is not just that a fish lives *in* the water and a bird *in* the air, but that the characteristic functions of these animals are what they are because of the special way in which water and air enter into their respective activities.[1]

The role an environment plays in the life of an organism is powerful and important. Organisms that need oxygen will not live long in a cave filled with carbon dioxide, so the role of oxygen would seem to be crucial in some contexts and to some measure.

As organisms respond to the conditions of their world, some responses are definitely experienced as more urgent than others. Some habits of living are simply not optional, such as breathing. These are the brute facts. We could thus say that such urgent habits involve biological *drives*. But the use of the word "drive" to define the relationship between biological and environmental structures in an organism has historically been laden with excess weight. We need merely say that reproduction, eating, drinking, and so on are regular parts of human interaction with the world, and that these habits are very important. However, the more subtle—and important—point is that human and animal organisms develop not only behaviors, but also biological *structures*, in relationship to the demands of an environment. A heart or lung, as a structure, relates part of an organism to part of the resources in an environment, providing the possibility for some flexibility and stability in that relationship. These structures become a part of the environment.

The genes of an organism are another type of structure; they encode the organic relationship between an organism, its progeny, and its environment. Genes inscribe instructions that help parts of an organism to implement methods for utilizing and coping with natural resources. Genes are merely one way of describing the complex metabolic interaction between organism and environment. They are not "prior" to the organism in a physical or philosophical sense. At conception, genes inscribe the ways in which the progeny of an organism have related to components of their environment. This is accomplished as the genes within a sperm and an egg are merged into a new genetic pattern. The *genotype* is a word used to describe the resulting genetic code. Every organism is said to have such a genotype. The information in the genotype tells cells how to specialize, so that the specialized structures of a human or pig or lizard may be created. In the genotype are instructions that dictate exactly how proteins must be expressed in order that a heart or lung structure can exist and function.

However, the fetal organism is not identical to that set of instructions.

In addition, the genetic material of the organism is itself not merely a copy of the genotype. As the cells in a human fetus replicate, it grows larger and develops. Each replicated cell carries the same genetic information that was uniquely determined in conception. Soon, though, the fetus begins to take in resources (at first via the mother, then by breathing and eating on its own). In that process, the organism ingests and is exposed to *mutagens*. Mutagens, such as carcinogenic chemicals in our food, ultraviolet light from the sun, and ozone in our air, are so named because they are associated with genetic mutations. This interaction can change the DNA in particular cells of the human body.

Contact with a mutagen will not turn a human being into a pig, nor will it convert a human heart into a cow liver, but it can result in "random" mutations that disrupt the DNA in a particular somatic or germ-line cell. While most of the DNA in an organism will remain stable after cells have mutated, there are many somatic and germ-line mutagens, and many phenotypic manifestations of their interaction with an organism's DNA. Many human cells contain DNA that has been slightly or dramatically altered through interaction with mutagens. Thus the organism's DNA, as well as phenotype, is modified in its environment. The majority of the DNA will remain stable (thus we can coherently speak of "genotypes"), but particular DNA can be modified by the intake of resources under conditions of the environment. As biological structures, genes react to and are changed by experiences of temperature, humidity, nutrition, smells, sights, and sounds. Genes are thus one important level of relation between organism and environment. Put simply, genes inscribe biological habits of relation between the organism and its environment at the molecular level, some of which are more stable and heritable (genotype) than others (bits of DNA in particular cells).

Genetic diversity plays an important role in the continuous developing relationship between each organism and its world. For every human trait, genes and environment are brought into a complex relationship. Genes structure a particular part of an organism's metabolism, enabling complex structures to undertake disparate tasks. Many genetic processes are at work in even the apparently simple human organs—a heart must be able to use energy in a particular way over a long period of time, and its tissues are subject to large stresses without subsequent relaxation. The trait "circulation" is defined to include an even more complex matrix of interactions. Beyond the heart, there is a maze of veins and arteries, and a filtration and oxygenation system. There is the blood itself and its constituent components. All of these systems in turn require a fancy array of metabolic interactions, and nutritional support for these metabolisms requires an environment rich in vitamins, minerals, proteins, and water. Genes effectively structure the metabolism of nutrients in such a way that circulation is possible within an organism. Yet genes are not static units, describing a single mode of relation between the organism and the world. Genes *alone* are not sufficient to cause traits. The identification of a trait is the identification of complex interrelations of genetic instructions and their environments, instantiated in particular organisms or populations of organisms.

Nonetheless, a few diseases, such as Huntington's chorea, have been identified with one or two genes that remain stable across virtually the entire population at all ages. Even in these cases, the etiology of the disease may be causally unrelated to the gene for the disease. And such single-gene diseases are extraordinarily rare; Huntington's chorea occurs once in every ten thousand live births. Yet it is on this model of simple biological correspondence that the whole Human Genome Project is built: complex human traits are taken to have one or two genes that remain stable across time and population. As we uncover diseases and traits that have markers, we must be careful to assess the markers carefully: Are they the *causes* of diseases? Do particular markers serve multiple purposes? Do they mark different diseases in different people? The marker for sickle cell anemia also appears to mark a resistance to malaria, and it may be that many genes have multiple functions. A genetic test for metabolic disorders turns out later to be a test for Alzheimer's as well.

The key to effective examination of hereditary etiology for diseases, and to the genetic "basis" for traits, is *context.* Apparently solid genetic relationships between populations of organisms and particular environments may be unstable, and even when such relationships are stable they may be more complex than it would at first appear. Cautious and experimental progress is in order, and as much attention must be paid to the definition of traits as to the location of markers. Moreover, we need to take much more seriously the undergirding components that fuel genetics: metabolism and nutrients.[2]

From this more pragmatic perspective, Dewey had much to say about the prospects for isolation of simple biological causes. First, he emphasized that processes that surround us are continuous, making it difficult to single out simple causes. Conditions change, and organisms do not stop to be measured. Second, he emphasized the critical importance of the *context* of measurement. Reductionist biological methods, which attempt to find genetic causes rather than genetic *and environmental* relationships, are often unable to produce good indices of causal relationships among genes and traits. Third, within the contextual and continuous setting of science, Dewey shows that the search for a particular cause is really a *selective emphasis* on a particular purpose—namely, to bring about or prevent some happening. When many theorists call something X the cause of something else Y, they wrench these events out of their natural contexts, they overlook the continuity of the process of which X and Y are parts, and they fail to acknowledge the purpose involved in focusing on these specific events.

The metabolic processes of living take place in a context. Organisms live in societies, which are as ritualized and complex as the processes inside the human body. Biology thus always and constantly meets culture, as genes are expressed always within environmental contingencies encountered by organisms within their cultural rituals.

Biology and society are connected in the most intimate way. Yet most people believe that the laboratory is a special zone of objective discovery. The perception is that scientists objectively measure the dimensions and workings of molecular life. Thus it is assumed (even by many scientists) that opinions about the social purposes of research are welcome only before and after we stop working on the data. Science, it has been held, is the realm of fact finding, not the place for social debate. Scientists produce the objective facts that we then evaluate as cultural critics. The facts are entered into evidence, and assessed from the perspective of what we want.

However, if you take a look inside a real lab or any of the institutions that support it, you quickly see that the "objective scientist" is a hero in an old myth. Contemporary decisions about which hypothesis to advance, which methodology to use, how to conduct observation, and most importantly how to conceptualize results and conclusions, are directly influenced by the values held by both the scientist and the institutions supporting the science. The decision to be rigorous and objective in observing an experiment, which we might traditionally think of as value-free, is actually a decision to do research in the way most likely to *work* given what we know about what has worked for scientists in the past—we have found that research conducted under certain strictures is more likely to produce results that we can put to use in solving problems. Moreover, the decision that we have "discovered" something, whether it be a gene, a quark, or a continent, is situated in terms of what opinions are held about the sufficient conditions for having discovered something worthy of scientific attention.

Thus while it is clearly important that the scientist be assiduous in observing a culture as it grows, the decision about whether to do that experiment in the first place, and how many times to replicate it, and what to say about what the data *mean* is always situated in a field of values.

The social problem with a strict separation of values and facts, and in particular with a separation of cultural values and biological facts, is that this separation is not present in our everyday lives or actions. Indeed, a separation of facts from values, or science from culture, is the antithesis of life in our culture. We see through technological glasses. The key is to learn to think about science and values as intimately intertwined. To do that, we turn to culture.

Culture

In the previous section, we saw that biology creates a matrix of resources and limits that texture the possibilities open to human life. The web of life, of which we are a part, creates both a context for our activities and a test that those activities must meet. Human beings, diverse and complex, take up biological resources at a variety of levels. Fortunately, we do not live all alone, attempting to create solitary means of flourishing in the *bios*. Our attempts to live in the biological world are always social.

In relation to the resources of this biological environment, our "organic structures" become "embedded in traditions, institutions, customs, and the purposes and beliefs they both carry and inspire."[3] We develop routines, which allow us (as biological creatures) to deal with the world of our experience. When the routines work, we repeat them and they become established as habits. Habits, or behaviors, become sedimented in culture. A habit might be transmitted by imitation, in textbooks, parental lectures, or laws. It can become foundational, as we grow increasingly sure of its ability to work. From birth, we are surrounded by people and animals acting out habits. In relation to these others, we begin to develop our own habits, under pressure from both biological limitations and cultural history. Human behavior is shot through with cultural routines and their meanings.

Culture is not just what we see at the art museum or the political rally. Culture is throughout the whole of our life story: "to indicate the full scope of cultural determination of the conduct of living, one would have to follow the behavior of an individual throughout at least a day. . . . The result would show how thoroughly saturated behavior is with conditions and factors that are of cultural origin and import."[4] This saturation, for example in language, relates our biology to our culture through activities:

> Even the neuro-muscular structures of individuals are modified through the influence of the cultural environment upon the activities performed. The acquisition of language . . . represents an incorporation within the physical structure of the human beings of the effects of cultural conditions . . . modifications wrought within the biological organism by the cultural environment.[5]

The biological structures studied by molecular genetics are constantly in relation to the habits of the organism and its environment.[6] Genes that get expressed in an organism are mediated by a world full of conflicting pressures. The cultural habits of a society embody our long-standing ways of dealing with and improving our biology. Thus biological facts come into relationship with culture the moment a new organism appears in the sphere of social life.[7]

We need to correct the idea that biology, or science, holds the key to all of the secrets about meaning in life. Biological investigation cannot describe the totality of human experience, because the values and purposes that give meaning even to the investigation of biological markers are not contained by simple units in the organism.[8] Value occurs as biological conditions issue in social and personal habits.[9] Similarly, though, the gene is not merely an instrument of social purposes. Its role in articulating one part of the organism's relation with its environment need not be denied or reduced to culture. Emphases on biology or culture are merely emphases; they do not reveal separate metaphysical realms of fact and value. Society emerges within biological parameters, yet also changes those conditions and thus transforms itself.

We also need to correct the idea that the scientist's character is somehow more likely to be value-neutral. The scientist is a human being, and the gathering of facts is a value-laden process. Geneticists are a part of a larger community of scientists, and members of institutions and families. It is not cynical to point out that, in addition to valuing human health, scientists value their careers. "There are straightforward economic and status rewards awaiting those who take part in the [Human Genome] Project. . . . Research scientists are not only involved in this struggle as academics. Among molecular biologists who are professors in universities, a large proportion are also principal scientists or principal stockholders in biotechnology companies."[10]

The pursuit of particular genes for illnesses is also, in most cases, the pursuit of the welfare of patients, and of the idols of the tribe and marketplace. In the case of genetic investigation involving traits such as criminal behavior, the institutional, social, and economic interests become very great.

Common Sense

Philosophical systems must be tested where the rubber meets the road. It is fine to have a comprehensive account of biology. It is equally nice to think programmatically about the meaning of social symbols and the community's pressures. But the usefulness of our approach to genetics must be tested in the context of ordinary people. Biology meets culture at the dinner table, in the reproductive context, and in real conversations about what research to conduct and when to develop a gene therapy. It is the field that Benjamin Franklin, perhaps our nation's most accomplished philosopher, called the sphere of common sense. In the field of common sense, we actually encounter things like assisted reproduction as a concrete option, and then we have a *problem* in our real lives. Our goal is to figure out enough of the biological and cultural context to enable us to do the *right* thing. We want to bring a sense of settlement, fulfillment, or *stability* to our biologically—and culturally—bound lives.

We don't always succeed. Common sense solutions to life's problems can be intelligent or not-so-intelligent. The less intelligent approaches produce a feeling of settlement, without resolving matters in a way that satisfactorily links our actions with our aims. Examples of such solutions include rationales for action based on "custom or tradition, obeying political authority, accepting some divine will, conforming to the wishes of the wealthy and powerful, resorting to partisan politics, and so on."[11] We might thus choose not to use genetic tests because we want to avoid "playing God," and feel great about the choice without actually understanding what it means at all. Such solutions are, for Dewey, "vestiges of a time . . . when the practice of knowing was in its infancy."[12]

The key to avoiding bad or unintelligent solutions to our actual problems is this: we must resist the tendency to explain away the "felt" aspects of the problem, as well as the tendency to apply abstract principles that might not apply to the problem at hand.

> In ordinary language, a problem must be *felt* before it can be stated. If the unique quality of the situation is *had* immediately, then there is something that regulates the selection and the weighing of observed facts and their conceptual ordering.[13]

Intelligent solutions to social problems begin with the feeling of a problem, then turn to work on the problem's basis in biology and culture. This is what we mean, philosophically, when we say over and over again that we must restate the ethical analysis of the biological and cultural dimensions of genetics in the context where our analysis *matters*. We have suggested an alternative way to reconstruct both biology and culture to take account of new genetic technologies. It is incumbent on us now to bring that reconstruction to the context where it matters. Just as the failures of current media discussions and cultural criticism about genetics are primarily attributable to a poor understanding of genetics and of social history, the test for us is whether the pragmatic account of genetics can return to the ethos where genetic interventions are used.

While the issues of biological reproduction "matter" for society at large, for science, for business, for industry, and for religion, the sphere within which our decisions about genetics are primarily made is that of the family, and particularly the institution of parenthood. So, as Jean Bethke Elshtain puts it, "we are back full circle, to concerns with the nature of human intimacy and the family."[14]

Commonsense Reproduction and the Idea of the Perfect Baby

Parenthood is the principal social institution concerned with reproduction. Numerous folkways (cultural habits) surround and give wisdom to parents as they make reproductive decisions. No argument for or against the appropriate use of genetic technologies has paid sufficient attention to parenthood, yet it is in exactly that context where some of the most important decisions will be made—and where biology and culture are expressed.

Hope defines the journey of parenthood. Aspirations of parents for themselves and their children create the context for reproduction. Hope for a parent's "perfect baby" is central to preparation for birth and parenthood; just as our society celebrates the marrying woman as "perfect bride," so too there is emphasis, as an intimate relationship with baby is undertaken, on the perfection of the child. "Perfect baby" is an icon of the journey of hope. More than hope, though, responsible parenting also involves *choices*. Choosing to *make* a baby involves a commitment to work to make life better for that baby.[15] We choose to make some of our hopes come true by participating in prenatal care. After birth, this hope suffuses our desire to make of our child a person of whom we could be proud, whom we can respect. Taken together, the choices and hopes of parents create a moral atmosphere in which our children dwell.

Yet these hopes and choices have a dark side. Too often the hopes of parents are unrealistic, misplaced, or lacking in foresight. Just as the image of a "perfect bride" has become, in our popular literature about weddings, a standard rather than an attitude of celebration, baby may be forced to attain rigid or foreign ideals, rather than to enjoy perfection of its own sort. How commonly the child of an athlete finds himself prefigured as "child athlete." One of our culture's common stories is that of the "first doctor in the family," forced from a life's dream of acting into medicine.[16] The choices of parents create a value system or ethos, and often that ethos can be too restrictive; it can become an oppressive standard that children are forced to meet. Our hopes must be mitigated by a readiness to elasticize, perhaps even untie, the child's connection to parental ambitions.

Though parenting involves hopes and choices that are profoundly important, many of the choices are not enunciated, or just seem to happen in the stream of daily experience. In some measure, this is appropriate. Parenting, as creation, is sexual and intimate—not the stuff of calculation. But even as the child grows older, it may remain unclear what impact our myriad choices will have on our children. Only some of the important choices of parenthood are lovingly thought out in advance—mostly, parents muddle through dozens of choices every day with little knowledge of which actions and choices will "register" in our children's moral and physical makeup. Parenthood is often characterized by a felt lack of control and understanding, a continuous struggle to keep up with changing children in a changing world. Choices and hopes never seem to work out perfectly.

That is at least in part what makes for the attractiveness of genetic interventions: they might enable parents to participate *scientifically and systematically* in the construction of their perfect baby. How might such systematic choices be made? Parents hope for healthy children and, if they can afford it, make choices (such as choosing prenatal care) to help "engineer" healthier babies. Genetic testing seems in this regard to offer the brightest hope for parents. Huntington's chorea and cystic fibrosis (and a few other rare diseases) might be isolated problems on the helix. Through genetic therapies, these disastrous, but genetically discrete, diseases might also soon be removed from the DNA of the fertilized egg or zygote. But, as we have seen, there are competing interests that make the effort to have healthy children more complex than simple choices might indicate. In Chapter 6, we will discuss the range of competing interests about health in the context of parental decision making, and isolate some possibilities and limits for curative genetic interventions.

Parents may also have the opportunity to make decisions about non-therapeutic modifications of their offspring. These decisions are not ontologically new; as we have seen, parents already make a variety of decisions that are hoped to enhance the quality of the embodiment of their children, from pre-natal care to breastfeeding to playing classical music near the womb. Our basic social commitment to education also belies the apparent newness of attempts at the modification of intelligence. Parents who would make decisions to use genetic enhancement, though, are subject to the complexities of reproductive decision making: we will see in Chapter 7 that most human traits depend on a complex array of socially and biologically intertwined characteristics. We do not need to look to Nazi Germany for the hazards of positive engineering: parents already make decisions that restrict the horizons of children. For example, though the child of an athlete would find the pressure of a father's prodding intolerable, the scientifically perfect infant, designed for the *purpose* of being a certain sort of child, would be pushed even more decisively (though less perceptibly to the child) toward a biologically predisposed character. The dark side of parental control is that choices and hopes can haunt the child: genetic engineering for the *purpose* of creating an exceptional child could be a radical extension of this dark side. In Chapter 7, we will discuss the range of options for positive engineering, suggesting the possibilities and dangers implicit in new choices for parents.

❧

Only by replacing determinist, reductionist thinking with a pragmatic recognition of the interaction of the biological and cultural matrices can we begin to develop coherent accounts of organic function, which both acknowledge the power of genetic structures and do not obscure their reciprocal, temporal relationships with the environment. Genes play an important role in the expression of traits, which in turn find their value and meaning in a cultural world. Genetic markers signal some traits that are relatively stable across environmental conditions, just as some environmental conditions set limits on organic life. The organism's condition is always a function of the interrelation of biological and cultural circumstances—neither of which "determines" the organism.

These relationships find their meaning in the lives of parents, who must make the most important choices about uses of the new technologies. Parental decision making involves hopes and choices, surrounded, tempered, and influenced by the biological and cultural contexts of life. By examining the ways these hopes and choices could be translated into specific options through reproductive technologies, we will recontextualize a range of choices for parents—each of whom wants a healthy, happy, "perfect" baby.

6

GENETIC APPROACHES TO FAMILY AND PUBLIC HEALTH

> Parents . . . seem surprised at their own helplessness in the face of the passion they come to feel for their children. We live and work with a divided consciousness. It is a beautiful enough shock to fall in love with another adult, to feel the possibility of unbearable sorrow at the loss of that other, essential personality, expressed just so, that particular touch. But [the new love of parenthood] . . . is of a different order. It is twinned love, all-absorbing, a blur of boundaries and messages. It is uncomfortably close to self-erasure, and in the face of it one's fat ambitions, desperations, private icons and urges fall away into a dreamlike *before* that haunts and forces itself into the presence with tough persistence.[1]

Pregnancy is scary and transformative; it is accompanied by a radical disruption of our ordinary lives. A wave of change, fear, and anticipation is palpable in physical, emotional ways. We become simultaneously hopeful and physically ill at ease. We are brought to the edge of our own frame of experience, where radical new possibilities become a part of our daily actions and words. The parent-to-be who is lucky enough to think of pregnancy as a positive development will still experience the loss of some possibilities, together with a banquet of new choices.

Interestingly, persons with terminal, long-term, or disfiguring illness frequently report similar experiences when they describe learning of their diagnosis.[2] For example, burn patient Donald Cowart changed his name to Dax in the face of his inability to continue the identity he had before his injury and treatments. In parenthood, a number of similarly radical changes in character are required of a woman—her body changes, she experiences new pains and fears and emotions, and she is surrounded by people who expect new things of her, including the miracle that she will have a living thing come out of her body.

Whether our pregnancy is something we have fought for, an accidental result of a chance encounter, or even an artificial insemination, at some point it begins to be a chosen journey toward parenthood. In its midst so many questions must be answered. Do we have sufficient insurance? Will we be able to deal with the constraints of parenthood, including changes in careers, plans, finances, family and work relationships, and space requirements? Confronting the loss of some possibilities can cause enormous strain. And there is the ever-present question: will the baby be okay? During these initial days, very little bonding with a fetus takes place. The pregnancy might be identified only by its attendant sickness because the fetus seems to have as its sole means of communication the capacity to make the expectant mother hungry, sick, tired, and moody. So, as Hans Jonas points out, the fetus is initially felt as a kind of parental obligation to the future. The early opportunities to relate to the fetus are distant and impersonal: one should eat grains, drink lots of water, take pills, and abstain from alcohol and drugs. Parents ache to know and to do more.

All of these changes are, of course, accompanied by myriad joys and possibilities. A baby promises a whole new world of activities and development and fun, and brings rewarding new responsibilities for the judicious parent. For the lucky ones with resources sufficient to the task, the world is asplendor with babies and infant accouterments. And for every expectant parent there are choices to make, for which we have been prepared by our culture and its iterations of the "biologically" or at least "medically" responsible parent.

Among the cardinal concerns for the expectant mother and father are those of health. This is certainly appropriate, writ large. Yet the aim of attaining a healthy baby involves complex decisions and is attended by diverse and incompatible social imagery about the meaning of health. So, as Molly Ivins sagely noted, where parenting is concerned the devil is in the details. All of the decisions about baby's health must be made in a context-bound, ongoing, and experimental way—decisions at eight weeks gestational age are different from decisions at age three. And while our culture has dozens of manuals for the expectant parent that deal with preparing baby's room and taking care of oneself as a mother, we have only begun to think about how to counsel parents for diagnostic decision making—let alone in vivo therapy—involving genes. It is time to set such an agenda.

Health Choices

Presented with a dizzying array of choices, women and their partners today confront options that were not dreamed of twenty years ago. Parents are now presented with the opportunity to define an "acceptable child" with reference to standards they may not ever have contemplated, measured in terms of risks to populations, and with grave options attached. Ten years from now, the opportunity to test a fetus may be connected to the opportunity to treat it with gene therapy in the womb. But today, the choice of testing always comes attached to a more grim decision: what will we *do* if we find out that the fetus has Down's syndrome? What if it might get breast cancer?

Conversations about the use of new genetic technologies, in the form of genetic testing, come fairly quickly in the relationship between an expectant couple and the obstetrician, nurse practitioner, or in cases where such choices are anticipated in advance, genetic counselor. The histories of parents are laid out in evidence, and theories of genetic inheritance are put to work in determining whether testing and other protocols will be appropriate.

The history of both parents' illnesses, and illnesses of their families and ancestors, give clues about probable fetal anomalies. Heredity is dangerous—the grandmother's heart disease may be congenital, and the late-onset Alzheimer's disease present in the grandfather may be genetic as well. Some tests will be brought up by the OB-GYN, others may not be mentioned. There is little agreement among genetic counselors, let alone the general obstetrics community, about what information is owed to the expectant mother. Should every expectant mother know about every possible test, or only those the clinician believes are appropriate? There are no laws and there is no professional consensus to answer this question. You might get an obstetrician who volunteers information about a genetic test for Alzheimer's, or you might get one who thinks that test should not be used in the reproductive context to make reproductive decisions.

The reluctance of some clinicians to point out the entire panoply of genetic tests may be attributed to the danger associated with the procedures in genetic tests. They are risky for the fetus. Not terribly risky, but even a 1 percent chance of miscarriage is tremendous for many parents when weighed against any but the most urgent genetic test. Thus clinicians frequently counsel or indirectly pressure patients to think long and hard about the risks associated with these procedures. The clinician may harbor his or her own prejudices about what kinds of tests are worth the risk, and these may be smuggled into the clinical context.

Not all reproductive diagnosis, of course, is genetic or high-tech. At the low-technology end, if Sarah has had a previous miscarriage, steps will be taken to prevent another. Eating habits become the subject of long lessons about proper nutritional foundations for fetal growth. Nor is all intervention related to the woman. A mate's reluctance to clean the litterbox or carry heavy things must be overcome: ammonia fumes are dangerous for the fetus, and excessive stress on the abdominal muscles of the pregnant woman is to be avoided.

But much of the important process involving new genetic choices is indeed woman-centered, and comes in two primary forms, with another lurking around the corner.

Amniocentesis

Amniocentesis involves the introduction of a needle into the uterus between the tenth and sixteenth week of pregnancy. An amniotic fluid sample withdrawn from the uterus may be evaluated for the presence of genes or genetic markers related to Tay-Sachs, Down's syndrome, and numerous other illnesses. Gender may also be determined. The procedure carries a roughly 0.5 percent risk of miscarriage, but provides more detailed information about more fetal anomalies than blood test-based screening of the mother and/or father. It takes several weeks to receive results from the amniocentesis, usually during the eighteenth to twentieth week of pregnancy, by which time abortion is clinically more complex.

Chorionic Villus Sampling

Chorionic villus sampling, or CVS, produces the same results as amniocentesis, but much faster, usually within hours or a day. It involves the sampling of the chorion, an outer layer of the placenta. Recent data indicate that CVS carries a higher risk of miscarriage than does amniocentesis (5.5 percent). Fetal blood sampling, or *percutaneous umbilical blood sampling*, involves the insertion of a needle into the umbilical cord, where a sample of the fetal blood is taken. In either case, the information obtained is similar to that garnered by amniocentesis.

Circulating Fetal Cell Analysis

It may soon be possible to collect information about the fetus through fetal cells circulating in the maternal blood. This would allow risk-free genetic testing of the fetus, removing one bulwalk against the routine use of genetic tests in pregnancy. However, the technical assessment of such testing is still in early stages.

The implications of this test are enormous. In a subtle way, risk to the fetus, even if it was less than the risk entailed by a pregnant woman's walking the streets of New York or driving a car in Los Angeles, allowed physicians and counselors to subtly discourage the use of genetic tests where there was no family history. This decreases the chance of a false positive result. If everyone can have a risk-free test, will those without family histories of genetic disease begin to request testing? At what cost? How many abortions of misdiagnosed fetuses will result? Is the fetus entitled to a margin of error? Should there be social rules articulating the sphere of therapeutic abortion, or would that injure *Roe v. Wade*? It is likely this test will be available commercially long before the public even recognizes that there are moral issues associated with its use. For parents, it will also be difficult to rationalize the decision to use, or not to use, this test. Later we will discuss its moral implications for parents. First, though, we need to look at the social and cultural context within which this and the other two tests are available to parents and other institutions.

❧

Decision making about all of these diagnostic procedures is difficult. The typical physician or genetic counselor opts for a "nondirective" approach, in which suggestions about the risks and benefits are proffered in a "neutral" way, which can leave parents feeling icy and indecisive. Parents look for, often even ask for, recommendations that may not be forthcoming, and may settle for hints gleaned from a physician's tone or manner. On the other hand, a health care provider who pushes for or against testing can, by relieving the parents of the pressures of decision making, expose herself to enormous legal risk should her suggestion turn out badly, as well as take over decisions that might more appropriately be centered around the subtle values of patients, their families, and their culture.

Counselor and physician decisions about how to present testing options and results are, in part, the product of particular modes of physician education and beliefs concerning the tests. Physicians have their own ideas about what medicine means, and those translate into ideas about what to offer and how to present it. They also have cultural values. Some physicians are fundamentalist Christians. Others love technology and believe in Leroy Hood's proactive medical posture. Others hold that medicine is a business.

The stakes for any clinician-patient encounter about genetics are high enough. Even in the case of the more common hereditary illnesses, such as cystic fibrosis and Down's syndrome, it is quite difficult to arrive at an appropriate choice. Parents who have hoped for a child will find the choice of whether or not to test, with its attendant possibility of a decision about abortion, very difficult. Other parents, some of whom already have a child with Down's syndrome, report that all of their hopes for a coming infant hang on a negative Down's test. Such choices show human courage and frailty worn to its core. Yet parents must make these choices. And increasingly they will not make them as a small group in a room, physician and nurse with patient and family.

Economic pressures weigh on poor parents, in terms of both medical and parental decision making. When society provides no prenatal care to fifty million of its members, many of those without prenatal care will give birth to children with birth defects. The allocation of health care resources results in a modification of the health of the society, family, and economy. For many women there are no doctors and nurses in the room to help with genetic decisions, and in fact there are no genetic or other prenatal decisions, because there is no money. For women who make too much money for Medicaid but have too little for private insurance, prenatal genetic tests do not happen. And for those with Medicaid, prenatal care varies, with many states offering only a tiny amount of prenatal care for teen pregnancies or the indigent, and with the majority of states in transition to some kind of Medicaid managed care program, many of which will not pay for genetic tests or therapeutic abortion.

Those with insurance see new people in the room, or at least feel their presence. For them the pressure is of the opposite variety. Insurance companies, and employers, have indicated a willingness to integrate genetic testing into the normal battery of tests for preexisting conditions. This will affect parents both before and during pregnancy. Already there are several well-known cases in which women's genetic tests were used to deny insurance for fetuses (as unborn members of the insured's family) on the basis of a fetal "preexisting" condition.

The Ethical Options

What can genetic testing really do for parents and society? Is genetic testing really different from other kinds of tests? Who has the right to genetic information? Families and others have to decide what they will do in the event a genetic test turns up a disease gene. To make these decisions, parents and physicians will have to grapple with the meaning of health and illness as these figure into a family's values. More, though, parents need to have a broad sense of how genetic tests can be used and what they really mean, and need to be able to cut through the social hype.

For some parents of a fetus with a gene for a known disease, gene therapies of several kinds may now or soon be available, and abortion is always an option. But some parents will request tests for "conditions" that might or might not be a "disease." We must also think about what to do to anticipate this problem. But to get at the ethical options for genetic tests and gene therapy, we have to see the genetic test in its cultural and biological contexts. We have seen that too many parents, and certainly our society, come at genetic testing with a naive faith in genetic determinism and a general belief that medicine is getting better every day. We must take off the blinders.

The Social Context of Genetic Tests

Genetic diagnosis seems to offer parents new choices concerning the birth of their children. Genetic tests may help them choose a mate or decide to abort a fetus. And genetic screening of fetuses or potential parents might allow society to take a new interest in the elimination of birth defects. But these new choices have historical roots that parents should consider before making decisions and that society simply cannot ignore. These roots extend to a time long before eugenics or genetic tests. The beginning of "proactive" medicine and the roots of genetic testing are found in the very earliest uses of the *autopsy,* or postmortem examination and physical diagnosis and the changes in medical practice that surrounded these techniques.

The early nineteenth century saw profound changes in medical theory, practice, culture, and institutions. In many senses, the medicine of 1790 would have been recognizable to Hippocrates and Galen, and the medicine of 1900 recognizable to us today. In the early 1800s, first and most prominently in France, the use of tests and instruments began to replace the testimony of the patient in the encounter between patient and care provider.[3]

Tests, stethoscopes, and autopsies were most prominent at this time in the care given by eminent clinicians to the poor in hospitals. The hospital was also new in this era, and a century would elapse before it had transformed from its many roles as poorhouse, almshouse, and site of charity medical treatment to its present status as the modern temple of medical science in which all social classes seek the most advanced medical care. The wealthy were still cared for in their own homes, and would continue to be until the early twentieth century. Wealthier patients were still able to dictate their own treatment to a great extent, partly because of the incomplete professionalization of medicine in this era, and partly because the relationship of physician to wealthy patient resembled the patronage of artists and musicians more than it did the modern physician-patient relationship.

Doctors did not control the medical encounter with most patients until the end of the nineteenth century, and newly available tests, treatments, and diagnostic procedures were one of the ways they gained that control. These data also refined the methods of physician examination, as physicians increasingly rely on visual examination of patients' bodies to make diagnoses. As Bichat put it, "you can take notes at the bedside for twenty years and still be confused; open up a few bodies and all the obscurity will disappear."[4]

Twentieth-century medicine has with every passing day replaced more and more of "talking to" patients with "looking" at them. If you break a leg, it shows up clearly on an x-ray. If you have prolonged postaccidental trauma, physicians can scan your brain with magnetic resonance imaging equipment, looking for pictures of the size and location of the lesion from which you suffer. The age of the autopsy has changed the meaning of medicine forever. The average medical office visit lasts seven minutes and involves almost no conversation. Most of the time we spend with doctors involves having our picture taken.

These technologies have been enormously useful, hailed in society as magic bullets. It is difficult to overestimate the value of radiological pictures in the healing of bone fractures and in tumor diagnosis. Yet radiology has known its own excesses of hope: the x-ray was thought by many to promise new, radiological studies of intelligence, character, and strength. Despite the enormous advances in medicine's ability to diagnose and treat diseases through technologies of radiology, autopsy, and now genetic testing, excessive hopes and a devaluation of patient experience mar the success of such technologies.

What we have also lost is the age of *illness*, in which patient reports of symptoms were central to diagnosis, treatment, and conversations between clinicians and patients. The ill patient is *ill* because something interrupts her life. Illness gets its meaning from the ways in which a body stops operating or cooperating with the person, and even may begin to get in the way. It is the lived problem of being unable to do, think, or be in certain ways. To put it bluntly, illness is the full-bore description of the meaning of being sick from the perspective of biology, culture, and personal life. Illness has all but vanished from the modern clinical literature, replaced by a simple description of the biological components of disease, those elements of illness that can be observed *without talking to a person.*

Some identify this as the shift from medicine as an art to medicine as a science.[5] In part this is accurate: the critical skills of heuristic listening and interpretation have been largely replaced by a scientific relationship mediated through impersonal technologies. Increasing diagnostic power allows physicians to rely less on the features that characterize illness and more on the body "as scientific object."[6] Genetic tests are obviously the final stage of identifying disease as a clinical construct in the body of the patient. They are present even at birth, long before a patient feels anything. They have nothing to do, in fact, with the patient's report of anything. The use of genetic tests thus seems to be a calculative matter of pure biology and social priorities. Early on people in the eugenic movement began to speak of a gene "pool." Genetic tests allow clinicians to play the role of lifeguard.

An illness-based model of pathology assumes that what matters about the patient's condition is that it results in an in-ability, or dis-ability for the patient. Disease is defined in a book, while illness is defined by how it changes my life. We need some new skills to figure out how to reintegrate "what matters" with the possibility of the genetic test. And we need some plans to avoid allowing genetic tests to become part of a mindless policy of "eliminating" weakness and disease, as though those are traits completely detached from persons.

Reconstructing Genetic Tests

Parents and social institutions have grown to accept the modern medical reduction of illness to disease, and the determinist attempt to trace what is important about disease to the genes. *Education about genetic choices must reintegrate illness and disease.* This is the pragmatic turn, integrating biology and culture, common sense, and the feelings we associate with being sick.

We should in general note that physicians—including obstetricians—lose innumerable contexts for therapy when they rely on interventions that may be accomplished in a fifteen-minute clinic visit and with a prescription. National health care reform can take such concerns into account. If we must do without national health care reform, managed care organizations must begin to reward physicians on the basis of how "well" their patients feel, and on the basis of how much long-term illness is prevented, rather than on the basis of how much money the doctor saves per patient in a year. It isn't that we need to pay more for health care, instead we must allocate our health care investment more pragmatically. Physicians also need to be better trained in conversation, especially the art of listening for patients' complaints and discussing patients' values. Most Americans would be stunned at how little training in the physician-patient relationship most physicians actually receive. And most medical students today learn only a modicum of genetics and nothing at all about genetic tests or their implications for families and society.

The ethical agenda for genetic tests is twofold. First, genetic "causes" of disease *must* be understood in a completely different way by patients, physicians, insurers, and the government. Genetic tests can only test for contextual relationships between people's genes and their environment, and produce answers only in the form of probabilities for populations of people. Parents and physicians have been too easily seduced by the efficacy of some genetic tests for single gene disorders. *Most genetic tests are not very useful, and in many cases we know much less than the test results suggest.*

Genetic tests do not reveal what an organism *will be*, they merely show particular relationships between environmental conditions and enzyme production at the time of the test. Thus linkages between traits and genes are always specific to the environment in which any given organism exists. However, the development of genetic tests has included only minimum (in many cases no) controlling of environmental factors such as nutrition and temperature. Even under the best controlled conditions, genetic linkages to traits are always expressed in statistical probabilities, which do not apply directly to each individual patient: a 60 percent incidence of Alzheimer's disease in a tested population does not mean that there is a 60 percent chance that a particular individual with the gene will get the trait. Nor does the gene necessarily "cause" the trait. A trait is "caused" by the particular relationship between an organism and its environment. Yet we have done very little testing of the nutritional or other environmental co-factors that would make a genetic test more or less appropriate.

Second, while radiological and genetic tests, with their pictures of damaged bones and mutant genes, sometimes provide useful information, parental choices hinge on the way a particular gene will affect the quality of life for a child and family. We need a way of talking about which tests help parents and others to assess their lives in an appropriate way. There is more to this choice than simple "autonomy" of patients. Our biology and culture have much to contribute, thus the way to frame the reconstructed question is this: how can we use genetic tests as a part of ethical parenting?

What Can a Test Tell Us?

Victor McKusick's *Mendelian Inheritance in Man* lists over four thousand human traits, diseases, and conditions thought to have genetic markers.[7] The unearthing of a genetic marker occurs when a population that evidences a particular trait also shares information on one area of a chromosome in common. The marker denotes the chromosomal area that might be statistically associated with the occurrence of a disease. Markers may be used in narrowing the search for a gene whose actual expression is directly related to a trait. Even before a genetic causal relationship is found, tests may be developed to show whether patients have the marker. A test for a marker might be said to be similar to the insurance company table that links crash losses to automobile color and make. A red car might be disproportionately involved in accidents, but the connection between red color and accidents is not likely to be causal. Red may *mark* another trait, such as lust for speed. However, the conclusion that a person who drives a red car is more likely to have an accident than drivers of, say, blue cars is warranted only from a distant statistical perspective. In any case, we don't significantly reduce the chance that a wild driver will crash by repainting his vehicle.

Some causal links between chromosomal information and diseases have been clinically accepted. In these cases, the expression of specific genetic information correlates with a body process known to be characteristic of disease.[8] With some diseases, such as Down's syndrome and Trisomy 18, an extra chromosome is present in the child. The presence of this extra chromosome is functionally related to developmental abnormalities. Other diseases are linked to the absence of a copy of one chromosome. The XO female (Turner's syndrome) has only one copy of the X chromosome.

Many genetic diseases are single-gene abnormalities. As the name signals, single-gene disorders involve a single mutant allele, which is directly related to the clinical manifestations of a disease. Single-gene disorders may be autosomal dominant, autosomal recessive, or X-linked. Autosomal dominant disorders are expressed when one copy of a particular gene, or "allele," on the autosomal chromosomes (non-sex chromosomes) is present. Such disorders may thus be inherited from one parent. Examples include polycystic kidney disease and Huntington's chorea. Autosomal recessive disorders require that two abnormal alleles, one from each parent, be present. Thus both parents must carry the abnormally expressed gene. These diseases are often more severe than autosomal dominant diseases. Examples include cystic fibrosis, phenylketonuria (PKU), and Tay-Sachs disease. X-linked (dominant or recessive) disorders occur when the mutant allele is on the X chromosome. X-linked dominant disorders affect males more than females, and include oral-facial-digital syndromes and Xg blood groups. X-linked recessive disorders usually affect only males, and include Duchenne muscular dystrophy, ocular albinism, and hemophilia A&B.

What Is a Genetic Disease?

When a disease is said to be genetic, the description looks like this: an allele is encoded with "incorrect" information, which tells particular cells to manufacture the wrong enzyme or the wrong quantity of some enzyme. An imbalance results, which causes a disease. But this description of the etiology of "genetic" diseases is fundamentally flawed.

A good example of the problem is cystic fibrosis. A person with the most frequently expressed allele for cystic fibrosis, f508, "gets" the disease when the expression of that allele in the genome, under a variety of circumstances, produces relevant symptoms. The disease seems to occur in most patients where the f508 allele is present, regardless of environment and behavior. There is a link between the function of f508 and the symptoms of CF. The chloride channel, crucial to maintaining proper mucus consistency in the lungs, is modified by f508. Thus f508 is linked by strong statistical evidence to patients who have CF symptoms, and is linked by molecular evidence to the clinical manifestations of CF in the lungs and pancreas. Yet some 30 percent of CF patients do not have f508. Instead, they have one or more of hundreds of other alleles. Moreover, not all f508 patients have CF. How can we explain this discrepancy, and what does it mean in terms of genetic causality?

The answer is that while emphasis in diagnosing some diseases may be appropriately placed on hereditary information, that information is always in relationship with particular environments. The environment in which an organism lives, and from which its nutrition derives, actualizes different genetic information. Persons with unexpressed f508 may not live in environments in which that gene receives the sort of stimuli and nutrients that would produce the CF symptoms. Persons who have CF but do not have f508 must also have different linkages between hereditary information and their activities in the environment. Thus the gene cannot be said to "cause" CF, unless cause is defined as a frequent and stable correlation between some genetic information and particular symptoms. *Bottom line: the cause of a disease can never be independent or prior to the environment of the organism.*

The presence of a genetic cause for some trait, in addition, only refers to the relationship between some gene and a particular trait. Other genes may also have a role in the trait, as *part-cause*, as may particular environmental conditions. The causal gene may also have a relationship with other traits. Thus it can have other externalities, or *part-effects*, that may not be anticipated in research to establish its relationship with a particular trait. *While genetic causal relationships can be extraordinarily complex, part-cause and part-effect help us to highlight for parents and policy makers the interrelationship of causes and effects between where we live, what we eat, how we live, and a variety of genes.*

Recall from Chapter 5 that genes and the environment are not two hermetically sealed units, causing each other to execute predetermined instructions. The chromosomes are altered as they move through environments, and the environment is altered by the actions of organisms. The net effect of this is that while gene lesions—and thus genetic tests—are important in designating possible patterns of relationship between organisms and their environment, it is unclear for whom the test will be accurate and utterly unclear that the genes are responsible for diseases. Chromosomes are biological patterns, operating in different people under different circumstances. Expression of genotype is also variable under certain circumstances, and sometimes the allele that is correlated with a disease is also connected to unrelated but desirable traits—as is the case with sickle cell anemia. *We must take care to prevent the use of genetic tests in the general population until a clear account is given of why the tests are efficacious. Where genetic tests are only accidentally correlated with an illness, they should not be offered. Moreover, genetic tests such as the BRCA1 test for breast cancer should not be made available to expectant parents, or couples undergoing IVF, under any circumstances, without carefully controlled conversation about the meaning of the tests and their relative inaccuracy.*

Can We Use Illness as a Key to Monitor Testing?

In choosing arenas for genetic research, we must take care to think about our search for genes in the context of the choices that search will produce. Genetic determinism can lead us to idolize the DNA, and to hypostatize a particular method of diagnosis. Yet illness is not only a lesion or code, it is also a social experience. The emphasis on genetic choices has led us to think that parents will be "playing God" by choosing tests; we feel this way in part because we sense the time lag between such choices and the onset of symptoms. The key is to focus on the symptoms. We have to think of genetic tests as interventions into known diseases, and then to think about what kinds of roles we want to have in our children's lives. Do we want to eliminate breast cancer for our children? What cost is there for us in doing so? Will we begin to think of our children in terms of traits; are only the children without disease worth having? By focusing on a frank discussion of such questions, genetic counselors with courage will be able to recontextualize the testing decision.

When illnesses begin to be understood as diseases, or are reduced to genetic markers, we also tend to overlook the complexities in both biological inheritance and lived sickness. What kind of life can a person with CF expect? What sorts of commitments are required of parents and others? Parents are much more interested in these questions than in the distant and impersonal statistics that genetic counselors sometimes provide. Parents want to help their babies, and to bring them into a world of possibility and caring. So, when confronted with the issues of testing, parents want to know what it is like to have a baby with the genetic coding for a disease. They want to know whether they can handle the responsibilities, and whether a child is likely to have a life filled with pain. Weighed against the option of aborting a fetus with which they have begun to form a relationship, the parents need to know things that genetic testing alone cannot reveal. *Testing without counseling and consent should be illegal, whether on the part of families or other institutions.*

How Can We Reform the Practice of Medicine?

Parents want to have healthy babies. It is their first and most appropriate hope. Physicians who work with genetic information must share this goal. Given our conclusions about disease, illness, biology, and culture, several suggestions emerge that allow health care professionals and parents to work toward this shared goal. First, physicians must be trained to understand, and explain, diseases in their lived context. Fundamental medical education reform will be necessary for this task. Instead of working on slides and corpses during the first years of medical education, students must be placed in clinical conferences about patients—learning about illness in its lived biological context. They should begin immediately to perform examinations, and these must be central to discussions of pathology. *The point is that practicing medicine, with its diagnosis, talk, and work with patients, should not be seen as distinct from theoretical training in pathology and anatomy—practice is too frequently seen as a kind of technical afterthought.* The theory and practice of medicine must be purposefully integrated. The practical consequences of such training will manifest themselves in an increasing ability to talk with parents about the experiences of other parents whose children have particular congenital diseases. Thus, while physicians may move from reactive to proactive *diagnosis*, medicine will still have to react to parents, situations, and the history of illnesses in the community.

The second corrective relies on the first: physicians and health care professionals must place genetic information in its context, and must do so in comprehensible ways. Genetic information, typically revealed in a telephone call about the "positive" or "negative" status of a fetus for some anomaly, holds sway over the most difficult decisions of pregnancy. *We must reconsider the "physician-as-tool" model, in which the physician is encouraged to see herself as a provider of services, in effect selling genetic information and then performing procedures at the behest of parents.*

It is, of course, incumbent on physicians to present all available information about a disease and statistical problems in testing. But more, physicians must participate in these decisions as members of the community. The values of the obstetrician and geneticist will always suffuse her conclusions about data, procedures to offer, and the kind of life an ill child might expect. While an appropriate posture toward obstetrical medicine requires that the physician allow parents to make many choices, the choices always require the physician to *do* something. Thus *so-called "nondirective" genetic counseling, which requires the physician or genetic counselor to dispense neutral information about diseases without making evaluative commentary, is just as dangerous as paternalistic medicine. The physician is unavoidably engaged in the care of parents and patient, and must work toward disclosing all the information available, including his own opinions.*

Is There Ever a Duty to Abort?

Robyn Rowland correctly points out that patients with complex illnesses frequently report that their lives are worth living. However, referencing only this fact, she lambastes all of genetic testing as inherently committed to eugenics. It is a position shared by many. But parents clearly have a responsibility to protect their offspring from hereditary diseases that would inevitably cause a newborn great suffering followed quickly by death, as in the case of Trisomy 18. Infants with Trisomy 18 suffer greatly, and less than 10 percent live beyond the first year. For parents in Trisomy 18 pregnancies there is no miracle shot: no Trisomy 18 cases turn out "fine." It is difficult to articulate the suffering of such families. Defaulting to "nature" here is not a possibility: once we know the fetus has Trisomy 18 we must choose how we are to exercise our ethical responsibilities as parents.

The danger of contemporary discussion about genetics in society is the likelihood that narrow-minded positions such as Rowland's will cause us to make big policy mistakes. There can be no question that a couple who determines that their infant is sure to suffer and die incurs special responsibilities, and among those responsibilities may be one to abort. From time to time genetic testing will suggest a duty to abort.

What about Social Pressures?

So far, we have described the parental and medical options for genetic testing. If parents and physicians all had equal access to high-technology medicine, and all medical resources were distributed equally, we might reasonably contain the discussion to those two arenas of decision making.

But parents live in communities, and many are employees. They pay for insurance, and each has a different income. Some parents have extensive educational experiences, and come from families with long histories of support and nurture. Others grew up in tiny apartments, waiting in front of a television set for a single parent to come home from a long day. For some parents, childbirth comes after a drunken evening with a stranger. For others, months of expensive IVF precede the event.

For the parents with employer-based insurance, prenatal testing is likely to be an option. Others in our society, though, have poor or no insurance. Our society has chosen to pursue a system of corporate insurance, available only to those with resources, and a federal safety net for those at the very bottom of the economic ladder. The majority have insurance and are able to elect genetic testing. They are also able to refuse. The minority, upwards of fifty million people at any given point during the year, have no insurance at all. They can present themselves to emergency rooms, where they may wait in line for hours in order to see a physician concerning a pregnancy, but routine gynecological and obstetrical care is unavailable. It is important to see this as a social decision. Our society does not provide prenatal care for many of its female members. As a result, a disproportionate percentage of the uninsured give birth to children with disfiguring, crippling illnesses. That this occurs is probably not a function of malice toward the poor. Rather, it is a by-product of an insurance system that excludes high-risk, high-cost patients.[9]

During the past twenty years, insurance has shifted from a community-rating system, in which a particular community's losses are averaged and a rate is set for a geographical area, to a risk-based system. Risk- and experience-based systems "presume that it is fair to charge different prices, or to refuse to insure people entirely, if they need expensive health care."[10] Whereas under community systems, the moral assumption was that a community shared the risks of its members, the risk-based system's moral assumption is that it is unfair to have the healthy pay for the sick. Risk-based insurance solves the problem of scare resources by eliminating care for those who (1) cannot afford their own care in the first place, and (2) are likely to need a great deal of medical attention. It does this in two ways. First, those who apply for employment-contingent, or individual, insurance may be assigned a rate that corresponds to their condition. The diabetic might have a much higher annual health insurance premium than the nondiabetic. Second, the applicant for insurance might be turned down entirely on the basis of preexisting conditions, an examination, family history, an HIV test, or credit and criminal record reports.

Under the present system, parents with insurance can expect that their infant children will be covered by family policies. Those who carry non-employer-based insurance might expect that additional members of the family will be subject to the normal battery of tests and questions before health insurance is offered. However, even the employer-insured employee can expect that things will be changing rapidly. Employers, who at present offer voluntary genetic screening for hereditary ailments that might be triggered by a work environment, will make decisions about screening as well.

Many fear that the inclusion of genetic information in normal screening by institutions will result in unprecedented discrimination against individuals. Whole groups might be excluded from insurance coverage, as in the case of those suffering from sickle cell anemia. Those who have participated in research studies that reveal an allele correlated with Huntington's chorea might be penalized by later having to reveal the results on an insurance application. Underlying these concerns is the general sense that penalizing a person for heredity is unfair: the insured person can do nothing about genetic markers.[11] In an effort to combat genetic testing, two states have passed laws declaring a person's genetic information the sole, private property of that individual. Similar federal legislation is in committee.[12] At the extreme, it is feared that insurance companies might gradually provide incentives for in vitro reproduction, then regulate the embryos to be implanted.

What Should We Do about Genetic Discrimination?

Fears of unwarranted genetic discrimination are a really mixed bag. The idea that new genetic tests for insurance would test things that "aren't our fault" and the corollary claim that "we own our DNA" both rely on genetic determinism. The former claim, that genetic diseases are altogether beyond the patient's control, misconstrues the role of genes in disease. Drawing on genetic determinism, disease is attributed to a prior, fixed genetic blueprint, set into motion at birth. Yet we have seen that disease links human behaviors to environmental conditions, and that genetic code merely sediments that relationship in more or less stable ways. Thus genes do not "cause" diseases, and *patients who have diseases that are marked by genes are no more or less at "fault" than patients with HIV or high blood pressure or mental illness*. It can only be said that some diseases occur more frequently than others, and in relation to a wider range of behaviors. Genetic diseases are as much the result of behaviors as nongenetic diseases are. Moreover, we argued in Chapter 5 that notions of genetic "disposition" or "tendency" fall into a fallacious understanding of the gene-environment relationship, so it will also not do to suggest that a person is "not responsible for genetic tendencies."

The laws now passed under this "genes are not my fault" rationale (in twelve states including Connecticut and now New Jersey) are really bad news. While they seem to be an advance (after all, now people with genetic diseases won't be excluded from insurance!), left in the wake of these laws is more of our old habit of finding ways to blame those with *particular* diseases. We seek to blame the ill for a variety of reasons. In the case of heart disease, eating habits are involved. With HIV it is sexuality that is at stake. Blame provides a way of justifying the felt injustice of a sickness in the community. But it does not address the real problem, namely our human need to minister to the ill. Blame, particularly in the case of insurance, seems to obviate our responsibility to the others in the insurance pool: if they could have prevented a disease, then they should have—and if they didn't, we shouldn't have to pay for it. So, in the case of genes, the patient is supposed to be free of responsibility. This distance is said to exonerate the patient of blame for any genetically caused illnesses of a variety of types. However, the supposed distance between genes and patient does not exist. Like all diseases, patient and illness are in a tight relationship, in which behavior can never be irrelevant. *Blame is never helpful for any disease and we must pass a law eliminating preexisting condition exclusions* of any kind *rather than merely exonerating those with bad genes.*

Absent such reform, however, insurers are right to note that genetic testing is not fundamentally different from HIV testing or family history questions. In all three cases, tests reveal information that may or may not be causally related to disease. With all three, diagnosis might be made before a patient evidences any major symptoms. And in all three cases, testing can tend to stigmatize people, belonging for instance to an ethnic group or religion or age group.

The expansion of screening to include a fetus is, however, problematic. We noted that the designation "preexisting condition" might be applied to the condition of a fetus with a gene for a disease. However, this move depends on the fallacious assertion that the gene is the same as the disease. The fetus does not have a preexisting condition. It has nothing except the gene. And the gene is preexisting only in the sense that the fetus has not been born yet. This use of "preexisting" equivocates on the accepted meaning of "preexisting condition," without meeting its essential test, namely that the fetus has a disease prior to being insured. *Unless insurance companies are willing to exclude all prenatal care on the ground that the fetus is not covered, we can assume that from the moment of conception all fetuses are covered by their family insurance, if only as a part of covering the mother.*

Genetic screening may also work out to benefit the members of particular groups in disproportionate ways. For example, because higher-paying jobs are likely to carry routine genetic screening, individuals of certain ethnic backgrounds at higher statistical risk for G6PD deficiency and sickle cell trait could have less access to those jobs. However, this problem is already evidenced in screening for hypertension, which occurs with greater frequency in particular ethnic groups in America. Genetic screening reposes the question of how health care resources are properly to be distributed. *Again the issue in ethnic genetic screening is not about genes but about preexisting condition exclusions in general.*

How Does Genetic Testing Implicate the Family?

The screening process can also provide only limited information. Two copies of the mutation f508 are required (it is autosomal recessive)—one from each parent—in order that a CF child be conceived. If a job applicant carries one copy of the CF mutation, does an employer have the right to ask or require testing of the applicant's wife? In the absence of this information, data concerning the applicant are all but useless. If an applicant tests positive for an autosomal dominant allele, who has a right to the information? A long-lost daughter, who might carry the mutation as well, might request that the information be released *regardless* of the wishes of the applicant.

We are forced to pit our claim that genetic information is not intrinsically secret or the property of the person against the competing needs for privacy and information within a family. A pragmatic reframing of the issue requires that we recontextualize the entire problem. First, business and insurance must use accurate genetic tests in a reliable way.[13] Where individuals cannot be tested with reliability, they ought not to be tested at all. If tests do involve the family, or if results implicate family members, a strong cause for disclosure is immediately present. However, this obligation is not made special as a result of the "special powers" of genetic information. It obtains because parents, who hope for health in their children, also must be able to make choices to preserve that health wherever possible.

The Insurance Issue in Principle

Many of the problems present in genetic screening go to the problem of risk-based, actuarial insurance *in principle.* Where insurance and health care are distributed equally, or at least a bare minimum of prenatal and postnatal care is available to all citizens, genetic testing for exclusionary purposes ceases to matter. A central moral commitment is at stake: how will we understand the social problem of medicine? For medicine is *always* socialized, it is merely a matter of what kind of socialization we want to endorse: will we invest in social medicine through free markets or a government-directed system? Either way, medicine will take place in a social matrix, in which the allocation of health care resources is a social decision. It is because we forget that medicine is socialized that we make such strange allocation decisions: any underprivileged patient who is sick can show up in an emergency room and expect to be treated, regardless of ability to pay. However, unless the patient is insured, she will not be able to get routine prenatal care, or be able to visit her personal physician. Resources that might have been spent on preventive health care for the poor are thus wasted on expensive invasive interventions for those same people, as their lack of preventive medicine manifests itself in chronic or acute illness. We think we avoided socialized medicine by refusing to insure the underprivileged person. Still we treat him in our emergency room. Only now we pay more. In genetic medicine this is the dilemma of thinking about genetic screening for insurance exclusions, rather than prenatal care for all. Genetic tests to screen out adults do not pay in the long run, as those who are eliminated from the insurance pool show up in hospital emergency rooms. Even in the short term, screens may cost much more than they are worth to insurers,[14] both because tests are expensive and because the concomitant elimination of surcharges associated with the diffuse possibility of a genetic disease in the population would disproportionately decrease premiums. I can't charge everybody for the low risk of Huntington's if I have eliminated Huntington's from the risk pool entirely. *Put bluntly, eliminate the gamble and insurance becomes less profitable.*

It is a mistake to frame the question of insurance or business genetic testing as an either/or, threshold-of-doom proposition. Decisions must be made gradually and experimentally, and as information produces new situations in which parents and children are exposed to new problems, the impetus to reconstruct insurance, while preserving incentives, will be palpable. Insurance is not just a simple business proposition; it requires us to reconstruct the interpenetrative health care relationships in which we all participate. As Tom Murray notes,

> Insurability is the set of policy decisions by insurers about whom to accept. It is not a trait, but a concept of membership. . . . Treated as a scientific fact about individuals, the notion of insurability disguises fundamentally political decisions about membership in a community of mutual responsibility.[15]

The Social Context of Gene Therapy

Beyond the techniques and technologies of genetic testing, abortion, and IVF lies the promise of *gene therapy*, curing hereditary diseases with pinpoint accuracy and finality. For many, gene therapy is the ultimate goal of the Human Genome Project. Faith in gene therapies among lay people is buttressed by a belief in genetic determinism: if we can isolate a disease-causing gene, we should be able to repair that gene—if not for ourselves, then for future generations.[16] Most of the use of genes in therapy, in fact, has not been an attempt to do anything like this. No attempts are under way to "rewrite" patients' genetic codes. And most of the use of genetics for therapy is completely unrelated to changing patients' DNA.

Reconstructing Genetic Research and Gene Therapy

Before we discuss the possibilities that use of genes in therapies may hold, we have to set those possibilities in context. First, the human gene therapies that we will talk about are predicated on extensive, expensive research. The Human Genome Project costs in excess of $200 million per year, and millions more in NIH funding are committed to clinical and lab trials of genetic therapies. This public funding would cover the cost of prenatal care for most of the infants on the North American continent. Because the presence or absence of prenatal care is so strongly correlated with pre-term births and birth defects, the use of funding in the latter way would cut annual medical costs substantially—one infant can easily run up $1 million in neonatal intensive care costs, almost all of which is absorbed by the hospital and passed on to paying customers. As a result, it might be that the immediate results of reallocation of genetic research funding would have a greater immediate effectiveness in decreasing the incidence of birth defects and other long-term illnesses.

However, social choices about the allocation of health care dollars are driven by a variety of forces. Genetic technologies are in demand, primarily, by paying, insured patients, who are represented in this case by powerful physicians working in academic medical centers and by pharmaceutical corporations and biotechnology companies. Medical and scientific establishments have a reciprocal interest in producing results that will be acceptable to paying patients, university trustees, and other institutional representatives of market forces.[17] And there are the higher callings: physicians and scientists want to find new and better ways to cure illness and to advance knowledge; however, the ways in which knowledge is advanced and illness is cured are textured by cultural commitments to particular methods for obtaining progress. Just as illness is social, our methods for obtaining cures are subject to our social and institutional standards for what counts as satisfactory progress. And these standards can seem odd, even striking, when we examine them against an even *slightly* wider context. Parents with insurance are able to visit an obstetrician and get adequate testing and prenatal care, and when appropriate they will be referred by their highly trained specialist to university hospitals, where they may have the opportunity to participate in gene therapy trials. Meanwhile, across town, an underprivileged patient may not see an obstetrician except by referral from the emergency room, and she will be provided with esoteric therapies (such as genetic therapies) only if she happens to go into labor while close to a university tertiary care facility that has Medicaid-approved clinical trials of an appropriate genetic therapy in progress. These two parents inhabit different worlds, in which parental options are, in part, a function of income. There is fundamental inequality in their options, and the Human Genome Project does nothing to narrow the gap between high-technology medicine for the well insured or the wealthy and triage-based medicine for the poor. It does not require that we think ten years into the future to imagine this inequality. The budgets for the genome project and, among other things, poverty medicine are set right now.

The answer to this dilemma, however, is not to pit the wealthy against the poor, nor to pit the genome project against poverty medicine. This is, however, the tack taken by Jeremy Rifkin and Robyn Rowland. The incoherence of directing research funds away from the genome project to prenatal care is apparent at once. First, it fails to acknowledge the importance of genetic research for a variety of collateral medical endeavors, including cancer and HIV research. Second, it treats only one symptom of the larger problem of the allocation of scarce resources. Third, it fails to target particular problems with the genome project's focus, settling instead for a hazy attack on genetics generally. Allocation of funding for human genetic research must conduce to the long-term achievement of social justice, but to make it meet this test, we must be careful in our assessments of need and priority.

In order to avoid empty comparisons between genetic research and other kinds of medicine, we have to shift our focus. Genetics and the genome project are actually dozens of research teams working on hundreds of different projects in different ways. What unites many of these researchers is a desire to heal, a faith in genetic determinism, and access to a massive research budget. We must focus—in funding and in actual research—on the particular purposes of each researcher. Our "genetic" goals are variegated: to find genes for diseases, and to create gene therapies for a variety of different diseases, only some of which are genetic in origin. We have to focus on each of these *particular purposes*, authorizing funding on the basis of the particular importance of each goal. Our social goals for medicine are equally variegated. Many believe that all should receive basic health care coverage. Many, such as Robert Nozick, hold that medicine should operate on the basis of a market economy. Debates take place in open hearings at the National Institutes of Health and in Congress about the proportion of research to public health dollars. However, the Human Genome Project, because it is a distinct NIH line-item budget, is exempt from these discussions. Reauthorization hearings for the genome project are a separate matter.

Those who apply to conduct gene mapping work, or to do gene therapies, and the genome project as a whole, should not be exempt from external peer-review. That genome sequencers do not have to run their work by scientists who are not involved with the project is part and parcel of the "superproject" principle, applied as well to the Apollo Mission, that governs its existence. Like the effort to get to the moon, it was brought into existence as a kind of special, national priority. Its privileged status depends on a social faith, expressed through a congressional vote, that genetics holds exceptional promise for solving our ills and needs special, accelerated funding. This was an enormous organizational mistake, for several reasons. First, to make a biomedical approach "special" violates the principle of peer-review science; a protected class of research is created, which is not subject to comparative merit review and does not have to compete for funding with other scientific endeavors. It is taken as a given that work on sequencing the genome is necessarily meritorious, and that grant applications should merely sort out who can do the best and fastest job of mapping and sequencing. By funding the genome project as a separate part of NIH, Congress gave its imprimatur to the notion that genetic research is special. The rationale for this move was that the only sure way to fund the project to completion was to set it apart from other research. But this move begs all of the important scientific and philosophical questions about whether its completion is desirable or possible and at what cost.

To put it country simple, the genome project should be reduced to its constituent parts, each of which should receive evaluation alongside all other genetic research. There simply is no justification in the present research and public health climate for a separate and unassailable genome project in the United States or elsewhere. The search for a genetic cure for cancer must be placed alongside other cure research and evaluated by comparison with a pool of applications to see which offers the most cost-effective approach to which pressing medical problem. Genetic mapping must be evaluated alongside other high-priority activities. It may be objected that our approach ignores the scientific value of mapping, as well as the likelihood of discovering accidental markers in the process. However, the scientific value of mapping has to be cashed out in a complex society with differential needs. Genetic mapping must be continually reevaluated in the context of a range of other social goals, and the attempt to make manifest its relatedness to illness must be plain. Mapping for its own sake has a value, but that value might not be sufficient to outweigh a competing application for rural medical upgrades or prenatal care. Directly or indirectly, these projects should have to compete with each other. In cases where mapping for its own sake is compared with nongenetic cancer research, the possibility of discovering markers-down-the-road might not be as valuable for medicine, society, or science.

This point is not only administrative. It is ethical. Our national commitment to a separate genome project bespeaks our love of high technology and the bedazzled way in which we treat its fruits. We must not allow our idolatry of technology and its attendant "Idols of the Body" to dissuade us from the project of curing illness in the most efficient and expedient ways. Rifkin and Rowland are wrong to insist that genetics is sui generis detrimental to poverty medicine. In order to hold such a view, they again have to rely on the strong claim that authorization of any such research is a slippery slope to the eugenics that may flow from the genome project. Instead, we have to hold each researcher accountable in the normal way, so that applicable discussions of funding priorities obtain for genetic research—and so that genetic research is always linked to its social fruits in curative treatment. We do not need a Human Genome Project to have many of its benefits. It is only necessary that these benefits meet the individual tests applied to all biomedical research and are fundable given NIH priorities.

What Can Gene Therapy Do?

The first wave of gene therapies has met with a great deal of enthusiasm from patients' families and the public. Though still in clinical trials, and in the absence of proof that any of the somatic cell therapies will ever work, gene therapies have been well publicized and thousands have applied to be included in trials. Gene therapies are carefully distinguished in the publicity from genetic engineering or genetic interventions, though in practice there is no distinction to be made. "Engineering," "therapies," and "interventions" all refer to the same thing. Gene therapy is merely the use of one or another kind of genetic modification by physicians. The various genetic therapies do vary dramatically, however. Their mode of operation, efficaciousness, and the purposes involved in their use distinguish them from each other.

As outlined in Chapter 1, genetic therapies are of four different kinds: the creation of compounds through gene splicing, invasive euphenic intervention, somatic cell intervention, and germ-line intervention.[18] Different therapies that use genetic materials or modify human genetic makeup accomplish different purposes; they are all working as well on different problems, for which varied nongenetic therapies also are available.

Gene Splicing

The creation of compounds involves splicing into bacterial cells some instruction for the creation of a human enzyme, hormone, antigen, or other protein that the body fails to produce. The bacteria reproduce and provide a quantity of the needed compound. Engineering of this kind has replaced animal insulin for many patients. These therapies are "ex vivo," meaning that they introduce an amount of new genetically engineered material into the body. The patient's cells are not required for the creation of the compound, because human insulin, for example, is not patient-specific. Thus the patient's actual genetic information is not removed, altered, cultured in bacteria, or reinjected.

There seems to be no reason to object to the use of such compounds. They do not alter the DNA of the human who donates the original DNA from which insulin is cloned, nor the DNA of the human who receives the injection. Yet genetic technologies brought us to this technology, which involves the somewhat radical leap to "cloning" the particular information for some human enzyme or compound.[19] How will parents who agree with Jeremy Rifkin or Paul Ramsey square their concerns about religion and natural balance with the realities of this noninvasive, nonmodifying, yet genetic therapy? Of course they cannot. Yet it is in the context of such decisions that the concerns of Rifkin and Ramsey must issue in responsible actions. If God gave us our biology, which we ought not to modify, can we at least make copies of small bits of that biology and use them, without changing our divine information? If the natural balance is upset by genetic modifications, what ought we to do about the possibility of restoring a natural balance in a diabetic's chemistry through genetic modification? How much more unnatural is the use of Humulin (genetically engineered human insulin) than the slaughter of thousands of cows for insulin or the death of unmedicated diabetics?

A pragmatic approach to therapies that involve the creation of compounds must set Humulin against the other clinical options. It seems to work better in most cases, and is not prohibitively expensive. There is no slippery slope involved here either, unless we begin to accept the claims that the creation of genetically engineered enzymes will lead to genetically cloned brains and genetic pacification compounds. The essential issues concerning such engineering are those of cost: is it more effective and affordable than other diabetic therapies? There is a critical distinction between cloning an insulin-producing cell and cloning a person: one does not lead to the other. Engineering these compounds neither advances nor retards the entirely different, much more problematic research concerning the reconstitution of "personhood" through genetic cloning.

Invasive Euphenic Intervention

Invasive euphenic engineering is distinguished here from the creation of compounds through engineering only to show that particular kinds of therapies differ in their delivery mechanisms. As we noted in Chapter 1, in invasive euphenic engineering no genetic material is removed from the patient, and the patient's DNA is not the subject of an "editing procedure." Genetically engineered compounds are used in pinpoint surgical procedures, such as the University of Iowa's modified herpes infection, which enables the treatment of some cancers.[20] The technique, again, involves sophisticated engineering of viruses, enzymes, or bacteria, which are injected into the patient. And an additional step may involve further invasive treatments, as in the case of the Iowa experiments: the infected cancerous cells are destroyed with a ganciclovir injection. The techniques, though, draw on the classical treatment methodologies for cancer and other diseases. The method is to destroy cancerous cells, rather than neutralize or reverse their mutations through modification of genes.

There are some additional risks involved in these invasive therapies, some of which deserve attention here. First, there is the risk that modified viruses will revert to their "wild type" and become infectious. Modifications of viral cells occasionally have this outcome, a result of the imperfect process of cloning new cells in the lab. Future generations of the cells, if they revert, could be dangerous and infectious. The patient with brain cancer could end up with both herpes and brain cancer. A second risk is related to the first: modified viral cells could conceivably mutate into dangerous new strains of a virus, which would be difficult to contain or treat. Both of these risks demand that scientists and clinicians work together on the development of invasive euphenic therapies.

Caution seems to be a wise posture toward the two varieties of compound-creation engineering. Their risks need to be studied, and in the process of granting approval for human-subjects trials, the National Institutes of Health must be carefully attentive to the emerging data on wild-type reversion. Given these risks, many potential genetic therapies may have to yield to safer alternatives. Yet the progress in such modifications of viruses and compounds need not be cause for excessive fear or support for genetic determinism. The halcyon hopes of Jean Rostand for a superman are hardly advanced by the creation of Humulin. The similarities between compound-creation genetic therapies and current nongenetic therapies are such that fears of the commodification of nature or dehumanization of illness also need not enter into play.

Somatic Cell Intervention

Somatic cell therapies, such as the cystic fibrosis inhaler trials currently in progress in centers around the United States, involve the actual (and deliberate) modification of a patient's somatic (non-sex cell) genetic material. This simple description, however, is misleading. Gene therapies are of at least three different types. Gene insertion involves the simple insertion of one or more copies of the normal version of a gene into the chromosomes of a diseased cell. In this case, no attempt is made to correct the disease. The goal is to implant replicating versions of DNA that can produce whatever antigen, chemical, or hormone is necessary, in order to overcome a deficit or imbalance present in the diseased cells. In simple terms, the goal is to overwhelm the actions of cells with defective genes with the actions of cells with corrected genes. Such therapies simply internalize the delivery mechanism for engineered materials: insulin-producing cells are placed in the body, so that insulin injection is less necessary.

Gene modification would be the next step. Here, the goal is to modify much of the material in the patient's cells. Such therapy would at times be "in vivo," involving the patient's own cells. The cells would be extracted, modified, and reinserted. In most cases, though, "ex vivo" vector cells would be sufficient. For example, in the case of the CF inhaler therapy, a flu virus (adenovirus) is engineered to carry genetic material that, when a patient "catches the cold," will modify the cells that are infected with the virus. Until the patient is able to develop immunity to the cold, lung function is improved. Improvements are not permanent and cannot be passed on through the germ-line cells, and the risk is substantial that the virus will revert to wild type or just plain clog up the lungs of the already weak patient with cystic fibrosis.

The third variety of somatic cell genetic therapy would involve in vivo modification of diseased, or causative, cells, for example in fetuses in utero. Such therapies are entirely theoretical. Thus while the principle is fairly simple, its application may be much less so. The theory is that through some delivery mechanism, a patient's somatic material could be slowly modified to express a correct (nondiseased) allele. The patient would lose all expression for the disease, and thus cease to suffer from symptoms.

All three varieties of somatic cell therapy share some technical problems. The first is the vector problem. Put simply, the modified viruses used to get the new DNA into the cells of the patient have to date not been effective in transferring the genes. The viruses don't work in humans. The second problem is that even when better vectors establish good gene transfer rates, the body still eventually develops an immunity to the vector virus and the whole therapy is finished. Because no permanent change has been effected, the patient is not cured.

Germ-Line Intervention

David Suzuki and Peter Knudtson, the foremost social critics of gene therapy in the popular press, draw an ethical line in the sand at germ-line gene therapy (Principle No. 4 in their "principles of Genethics"). For them, issues relating to somatic cell therapies are merely practical, while the powerful germ cells hold all of the ethical import. We will shortly see that such claims about germ-line modifications rely on genetic determinism. At this point, we need only note that ethical issues in genetic engineering do not begin and end with the modification of future generations. Somatic cell modifications present ethical problems that suffuse practice: risks, costs, and benefits to humans are part and parcel of a pluralistic society. The additional fact that gene therapies are similar to other interventions in medicine also bears on the limitations and powers of genetics more generally.

Germ cell therapy is taken by Jonas, Suzuki, and Knudtson to herald the dangers of genetics.[21] The National Institutes of Health was so concerned with germ cell modification that it initially proposed adding a "check-off" box to its clinical trial application form for human gene therapies, which would have read: "Could this therapy be used to modify the germ cells?" An unsatisfactory response meant the death of the application. Though the box was never put into practice, the rationale for dropping it was that virtually no somatic cell therapy could meet the test. Genetic modification of germ cells is identical to the second and third variety of somatic cell modification, with the addition of the heritability of the modification. In other words, depending on the trait and its particular genotype, offspring of the patient would inherit the modification.

Jonas, Suzuki, and Knudtson are up in arms at the suggestion of germ-line therapy. What of the rights of future generations, asks Suzuki.[22] It is one thing to modify one's own cells, but to compromise the future is unfair. Jonas agrees, with a slightly different emphasis. His concern is that the freedom of experience—or, better, the freedom *to* experience—in life demands that children not be mere instruments of parental design. Germ-line therapies incur this problem in at least two different ways. First, they restrict the freedom or rights of a child that is actually born. The child is born "into the therapy," voiceless and unable to take back the actions of others concerning his health. Second, germ-line therapies restrict the kinds of children that can be born. Thus they overdetermine children.

This critique fails for two reasons. First, it fails to account for nongenetic activities by parents that have even greater impacts—including genetic impacts. Earlier, we noted that smoking during pregnancy is one such instance. Children cannot take back a whole set of parental actions, the sum of which necessarily creates the child in the first place. This leads us to a second problem with the critique. It depends on genetic determinism for its power. The only danger of germ-line engineering that is *unique* involves the possibility of utter control over the body that a child will inhabit. We have argued that no such power exists, particularly with reference to disease. In Chapter 7, we will see that the powers of genetic determination of social traits pale by comparison to the social power of parental activities.

There are problems with germ-line therapy. However, we cannot place enough stress on the fact that none of these problems has anything to do with the exaggerated claim that germ-line therapies will become eugenics, or decrease the value of life for patients with hereditary illness. Rowland also is wrong to fear that germ-line therapies will further ensure a genetic patriarchy. None of these claims, put plainly, has any basis in biological fact, which seems to make all of these claims somewhat morally moot. The similar harkening toward a slippery slope from germ-line engineering to genetic enhancement is based on our lack of wisdom—*yet if we lack wisdom, the correct way to find it is not to focus on the sensational possibilities.* The practical problems with germ-line engineering are sufficient to keep us engaged, and they embody the philosophical issues that are really important. Which diseases ought we to purge?[23] How do we assess the causal relevance of a genotype for a disease, and how do we assess the importance of the loss of its expression in the future? Accidental part-effects of gene therapies could accompany the intended removal of diseases. In addition, who can afford such therapies, and will their distribution be just?

Answers to these questions depend on our ability to see genetics as part of a range of options available to parents, society, and medical practice. Germ-line engineering is so expensive and risky that the implementation of it will depend on new and different advances. Moreover, in the present atmosphere of fear, in which modification of germ cells is taken to be meddling in sacrosanct territory, government and medicine are reluctant to appear to want to change the future. Two scientists working on an artificial cellular coating for use in IVF recently discovered the fruits of such effort, when they were suddenly catapulted into the spotlight as "human cloners" on the basis of a misreading of their work. The mere mention of a risk to the future frightens away scientists and clinicians.

It will be replied that these considerations of the value of germ-line gene therapy, while important, are not *ethical* in nature. *But it is exactly the point of a pragmatic analysis that the exigencies of a scientific and social ethos come to texture the possibilities for moral discourse.* So the possibility that germ-line engineering may not be seen as competitive with other therapeutic options is indeed revealing. We see that in order to make good decisions about germ-line therapy, it will need to be made clear that germ-line therapy does not uniquely influence the future and that its power is not prefigurative or determinative.

Pregnancy is frequently signaled by illness and a radical shift in orientation. Parents must learn to deal with new problems and to inhabit new moral relationships. All kinds of activities, from grocery shopping to listening to music, become decorated with parental hopes and choices, some more conscious than others, about healthy babies. Among these choices is the new possibility of genetic testing, which can sometimes yield information that helps parents to make better choices. Yet genetic testing presents theoretical and practical problems. To deal with these problems, we have to situate genetic testing in relationship to several transitions: the move from illness to disease in medical diagnosis, the increasing demand for information by insurance companies, and the problem of diagnostic uncertainty in genetic testing.

Parents who opt for gene therapies also face new choices, some of which are very expensive and risky. However, the suggestion that some or all gene therapies violate nature or God fails to help us make decisions. Similarly, differences between germ-line and somatic cell therapies turn out to be practical rather than ontological. In the place of the present theoretical scholarship, we have seen here that only a consideration of germ-line engineering in relation to other therapeutic possibilities will help make clear whether germ-line interventions are appropriate for particular illnesses.

7

THE NOT-SO-DEADLY SINS OF GENETIC ENHANCEMENT

Many critics of genetic research make reference to a slippery slope that begins with curing illnesses and ends in genetic modifications of appearance, intelligence, and character. It will be difficult to draw the line at negative engineering, Jeremy Rifkin and Robyn Rowland instruct, because the barriers to modification of DNA will have already been lifted. There will be new opportunities to use genetic testing and genetic therapies in nonmedical ways. Genes could be used to enhance appearance, intelligence, personality, and strength. Critics fear that these technologies could usher in a new eugenics, more dangerous than original eugenics by virtue of its greater genetic powers. Genetic optimists give the critics much to fear: Jean Rostand's superman, Shulamith Firestone's woman-centered socialism, and Brian Stableford's radical reinvention of human bodies all depend on genetic powers.

Yet genes did not usher in human hopes for self-improvement. It has long been an appropriate goal of societies, families, and individuals to improve their world. Parents frequently describe their desire to make a better world by having and raising children. American society devotes a great proportion of its resources to education, health care, and institutions whose purpose is to improve society along the lines of social hopes. Moreover, genetic optimists do not have sole purchase on the big dreams for improvement of human nature: Rifkin would have us improve our world by destroying much of our industrial production, and Rowland openly advocates a world in which technology is deemphasized. These too are social hopes, which would require tremendous change to accomplish.

In fact, genes may be the least effective means of advancing personal, familial, and social goals: we will see in this chapter that genetic engineering is unlikely to improve our character and intelligence. Nor is it likely, conversely, to destroy our human natures. Conventional social institutions, such as schools and churches, have a much more direct effect on our character and intelligence.

Finally, genetic modifications for "positive" purposes are distinguished from "negative" modifications in a way that few critics suspect. The difference is not, as is commonly held, that positive engineering advances social ideologies—while negative engineering merely cures the sick. Instead, the distinction between negative and positive engineering is in the *manner in which social improvement is attempted*. To see this distinction, we will examine the meaning of the two terms most commonly employed in "negative" or "curative" interventions: health and illness.

Healthy Babies?

Is there any real difference between curing illness and enhancing traits? To answer this question, we must pause to attend to the way in which the usage of "health" and "disease" has emerged in medical discourse and ordinary language. Health is commonly held to denote the absence of disease. This meaning of "health" thus orbits around the evolving catalog of diseases, so that when new diseases are identified, health is correspondingly limited to exclude the new sets of symptoms and etiologies. When an obstetrician refers to the "healthy baby,"[1] she is making a kind of "reverse diagnosis"—if the infant presents no symptoms of illness, it is healthy.

Though "health" and "disease" have their most sophisticated meanings in clinical discourse and practice, they also function in ordinary language. In ordinary usage, health has a somewhat larger scope. For example, the phrase "very healthy" might be used in regular conversation to describe the vitality of a fifty-year-old who, in addition to having no major illnesses, exercises quite a bit. When parents and others announce the birth of a "healthy baby boy," health can signify potential: the baby not only has no problematic illness, he also is, in his vulnerable state, potentially able to survive and prosper. When we refer to an acquaintance as healthy, we may mean that he or she seems energetic, fit, morally agreeable, or lacking in illness.

Distinctions between positive and negative genetic engineering rely on presumably settled operational definitions of health, the definitions purported to exist in the clinic. Yet parents who hope for a "healthy baby" may have more or less stringent requirements for health than their physician has; they may understand health in an entirely different way. So what *is* a "healthy" baby? Though ordinary language uses of "health" vary dramatically, the clinical identification of health as "freedom from disease" tells only half the story. Anthropologist Rayna Rapp at New York's New School for Social Research found differences between parents' and caregivers' perceptions of what counts as a healthy baby, differences only exacerbated for parents whose class and socioeconomic status differed markedly from that of their clinician.

In the clinic, disease is identified by its ability to cause bodily processes to deviate from "normal" operation.[2] A virus might cause a rapid acceleration of body temperature, difficulty in breathing, and dizziness. The virus is termed a disease because it causes symptoms that are not *normal* for a *healthy* human. Thus while clinical health identifies the absence of symptoms of illness, it also denotes the presence of dozens of "positive" traits, such as the capacity to breathe, pump blood, and behave normally. The designation of clinical norms for capacities has taken place during the 3,000-year history of Western medicine.[3] As medicine and science have grown increasingly specialized and prestigious, more and more norms have been developed, governing an increasing number of bodily interrelations. Taken together, a set of norms governing the interrelated capacities and processes of human bodies defines "health." As clinical diagnostic technologies continue to improve, the definition of health grows more and more narrow and sophisticated.

What we have termed "clinical health," then, involves having all of the capacities that are necessary to function within certain clinical parameters that pertain to various bodily functions. A healthy baby will have, among other things, healthy lungs. Healthy lungs can be identified in a number of ways, including a respiratory rate that falls within particular parameters. Similarly, a clinically healthy person has good eyesight. A person with healthy eyesight can read 1/4-inch letters on a chart at 20 feet. A healthy person is not mentally ill. The mentally healthy person can deal with grief, anger, and passion in appropriate ways. A clinically healthy person is within the clinical norms for all of the major areas of health, as roughly divided along the lines of the clinical specializations (ophthalmology, psychiatry, gynecology, etc.).

The lines of clinical specialization become important because health and disease are the business of physicians and nurses and the institutions they inhabit. Only if disease and health are involved is a physician consulted. Certain kinds of traits, ills, and experiences are outside the scope of medical discourse and medical practice: physicians do not treat avarice, atheism, or laziness, nor do clinical norms of health include courage, punctuality, or Christian faith. Just as religious institutions, academic institutions, business institutions, and government have their own spheres of concern, medicine deals with concerns about health and illness. Other institutions are reluctant to employ full-blown medical metaphors and protocols. A professor does not diagnose or inject her students, and pastors do not administer dialysis. While elements of the various professions suffuse each other, the basic language of medicine is its own, protected by culture and a long history of practical and discursive specialization. Taken together, the practice of medicine and the discourse of disease and health form a modern clinical paradigm.[4]

Much of the time, this conventional paradigm for understanding health and disease is helpful.[5] Fairly settled definitions of normalcy of pulse, respiratory rate, clarity of vision, and even color-vision are applied in routine ways during ordinary medical practice, with great success. These norms are the foundation of allopathic medicine, providing many of the basic diagnostic measures and enabling the measurement of success of therapies used by clinicians. Physicians develop habits for treating common conditions, which rely on the continuing workability of norms for disease and illness.

However, over the years more and more of our activities, from social behaviors to mating styles, have become open to clinical examination. This has resulted in the use of medical metaphors and protocols in dealing with a variety of human differences, not all of which are amenable to therapeutic interventions. Mental health has seen perhaps the greatest multiplication of behaviors to be added to the roster of sickness. Take, for example, the stability of personal identity. Through study of numerous patients, the diagnosis of schizophrenia has evolved to include a variety of conditions in which a person's identity is disrupted by illness. This "discovery" of dozens of varieties of schizophrenia carries with it a designation of *health*. To know when a person is schizophrenic, we must know when a person is *not*. In other words, it must be clear what it is to be "nonschizophrenic." In this subtle way, the stability of personal identity, which had not always been a "clinical" problem, came to fall within the province of medicine. The identification of any new mental illness carries with it the assumption that the behavior in question *belongs within the clinical sphere*. Eccentricities, forgetfulness, even lifestyles become not only desirable or undesirable, but also sick or healthy.

Yet in each case where human difference becomes the subject of the clinical constructs of disease and illness, there are questions that must be answered: Why should a patient's "depression" complaint—that life is unrewarding, unfulfilling, and sad—be reason to call a doctor? The depressed person could call on a philosopher, who might discuss with the (student) ways in which life can be enhanced by thoughtful choices. The person could call on a minister, who might provide the (parishioner) with some particularly inspiring theological text. The person could call on a biologist, who might show the (subject) the statistical correlations between seratonin uptake and depression. Each of these professionals offers a way of dealing with depression; gives a "whatness," a complex, socially sanctioned rationale, to the "thatness" first experienced by the person.[6] Social institutions grapple with different explanations of the "whatness" of depression, including philosophical, economic, biological, and medical explanations of prolonged unhappiness. At present, depression is typically thought to belong within the discourse of medicine. This way of talking about depression, though, has the interesting side effect of giving medical status to "nondepression," so that it is "healthy" to be "happy." We must remember that there are other views:

> . . . the members of [a New York therapy group] maintain that it is, above all, *society* that is desperately ill, that since we live in a world in which most people's values are hopelessly skewed, it is emphatically *not* the therapist's function to help patients adjust to the prevailing norms; indeed, the therapist should, if he is genuinely interested in leaving patients whole and at peace with themselves, encourage them to work, in any of a hundred ways, to alter society for the better.[7]

While medicine depends on the comprehensive clinical construct of health to treat the majority of conditions that afflict patients (from broken arms to myocardial infarctions), the expansion of medical territory has sometimes brought medical discourse to bear on complex social problems that medicine has great difficulty curing.[8] The Human Genome Project exacerbates this problem. Simple genetic linkage studies provide a research mechanism that can be utilized to link genetics to a variety of behaviors and activities. As we noted in Chapters 2 and 6, markers for "nonclinical" traits can then quickly issue in "diagnostic" tests, with a concomitant assumption that the "trait" is biological.[9] Moreover, powerful institutions can bring these tests quickly to the public, where parents and physicians are caught off guard by a flood of possible new tests, such as a test for homosexuality or aggressiveness.

The possibility that the notion of health will be expanded, so that physicians begin to treat social problems, is not a concern of Rifkin, Rowland, or Chicago social theorist Leon Kass. Their fear is that humans will try to improve their biology. Such an effort would be, Rifkin writes, categorically wrong. Rowland fears that positive engineering will remake nature in the image of men. David Suzuki and Peter Knudtson draw an absolute line between negative and positive engineering, as though the distinction were palpably clear. Only Hans Jonas casts his fear of positive engineering in terms of the inability of medical discourse to account for every part of human existence. It is this concern that plagues potential attempts to use medical technologies to improve the human condition: can genetic technologies, or medicine more generally, be an effective way of dealing with social problems? The image and disease-driven mode of perception may not have the tools to take over the roles of education, government, or even theology.

Medicine has the power to "medicalize" a broad range of human differences, and genetic technologies can facilitate the extension of this process. But should society allow genetic technologies to function in this way? To determine whether medicine—and particularly genetic medicine—is the best way of improving society while preserving individual freedoms and diversity, we will focus on the improvement of society as an aim, both general and medical. What is objectionable about genetic enhancement? Is it the hubris of human ambitions for improvement, the danger of freakish accidents, the expansion of medical territory, or the social consequences of attempts at improvement?

Reconstructing Enhancement

Lately it seems that a whole commercial culture and social conversation has grown up around "enhancements." Some are quite controversial: Prozac and other antidepressants have been increasingly reported to be performance-enhancers, and as Peter Kramer points out, they are even prescribed for that purpose.[10] Lawrence Diller highlighted an increase in enhancement-based rationale for use of the stimulant Ritalin, originally prescribed to combat attention deficit disorder.[11] Some enhancements barely raise our collective ire, such as the now well-established use of cosmetic surgery to modify appearance, the selection of offspring gender, or the sale of "genius" germ-line cells by one California sperm bank. Still others seem uncontroversial, or seem not to count as "enhancements" at all, such as the use of private schools, vaccinations, and vitamin supplements.

Because the coming opportunities for enhancement are so technologically advanced, we have been lulled into the impression that we are dealing with an entirely new phenomenon requiring new moral tools. In this chapter I argue that in at least one case of enhancement, that of the possible use of genetic technologies to enhance offspring, we are presented only with new bottles for the old wine. Our current cultural focus on the moral issues associated with "enhancement" depends, moreover, on a mixed-up understanding of "nature" and clinical "normalcy" that I term the "ontology of becoming." I make the case that reproductive genetic enhancement can best be understood within a wider range of other, more mundane parental decisions. The basic choices parents make about schools and nutrition and our ambitions for our offspring are inevitable and appropriate enhancement decisions. The question is not whether but how to enhance the lives and character of our children. All parental enhancements, I then argue, are subject to some dangers common to our cultural experiences of parenting. Paying attention to these takes us half the way to understanding why many genetic enhancements may turn out to be a mixed blessing indeed.

The Ontology of Becoming

What is the difference between engineering a child's genes and engineering a child's education? Why is private school not considered to be an enhancement technology, when Ritalin so frequently is? Two different ways of answering these questions have emerged. Each is part of what I term the "ontology of becoming," our cultural faith in a particular and deceptively clear description of the limits of human nature and the territory of clinical medicine. This faith has led us to suspect that we have clear purchase on what is and what is not enhancement. In order to understand why our culture selectively values some kinds of improvement of offspring, but selective disvalues others, we must first debunk the well-nigh entrenched idea that it is enhancement itself that is objectionable, rather than particular kinds or means of enhancement.

Our culture has great faith in the value, even necessity, of education. We spend billions on public education from cradle to grave. We hold that education is conventional and appropriate, unlike the enhancement of our offspring through drugs or genetics, because education allows children to fulfill their "innate capacities," without changing the child's "natural state." The best example of this is the lengthy cultural conversation about IQ. Intelligence tests and tests designed to yield an IQ score are predicated on the notion that we all have a fixed, immutable, biological capacity for thinking. No matter how much education you accrue, no matter how many books you write, your IQ should stay about the same. Physical change, such as brain disease, can change your IQ. But classes at Harvard should not. Hence the rationale for differentiating between two kinds of what, for lack of a better term, we can call "improvements in thinking." Education merely fills our buckets, and is thus appropriate development of our natural potential for intellectual flourishing. Modifying the bucket itself, though, would be a matter of enhancement. That many of us hold this faith is demonstrated by the fact that we don't blink when parents pour seventeen years and thousands of dollars into preparing kids for college, but balk at improving calculative speed through genetic enhancement or psychopharmacology. The former allows our children to "fulfill their potential," the latter is viewed as dangerous in part because it would modify potential itself.

Intelligence is only the most obvious example of our reliance in enhancement conversations on a hypostatized notion of the natural state, the first part of the ontology of becoming. Women are discouraged from cosmetic surgery and parents are discouraged from giving growth hormone to children on the grounds that it is unnatural and thus inappropriate to seek after a modification so significant as to constitute a move from one *kind* to another. Struggling for a rationale as to why cosmetic surgery is so different from, say, a new haircut, and thus might be dangerous, critical theorists point to the attempt by patients to change the kind of person they are.[12] Those who fear a slippery slope that begins with reproductive genetic testing for diseases and culminates in "an expectation of a perfect baby" cite a similar phenomenon: once parents have crossed over into modification of the natural state as a part of parenthood, it will be easy to modify that state for less meritworthy reasons.

There are two problems with this idea of natural kinds, though. On the genetics side, what looks like a stable matrix—this "genome" we're all trying to protect—is actually a complex matrix of interactions in three trillion cells, many of which have taken up small mutations. Walking in the sun ages your skin because it affects your genetic makeup. Radiation and the chemicals in the water effect changes in the germ-line and somatic cells. Even the air that you breathe is chock-full of ingredients that change the supposedly stable "blueprint" of genetics. There is simply no such thing as a genetic model that functions ideally in an ideal environment. Genes are affected by the environment, and the genome is our shorthand way of saying that we believe that most of the cells in our bodies have roughly the same genetic information, much of which we inherited. Yet much of the current bioethics scholarship has proceeded as though the genome is a "code of codes" rather than a relatively implastic inscription of hereditary information that is constantly modified in significant and not-so-significant ways.

The second problem is on the social side, and has to do with the cultural context of ideals like intelligence. Intelligence gets its cash value not from a sort of acontextual set of skills, but rather from the sum of skills a society happens to value. As Richard Lewontin points out, IQ tests actually ask questions like "Who was Wilkins McCawber? What is the meaning of sudiferous? What should a boy do if a girl hits him?"[13] Just as the seemingly objective categories in medicine, such as the definition of efficacious therapy, futile care and equipoise, really turn on social values, the meaning of intelligence and its supposedly objective root in "natural intellectual kinds" actually turn out to be a shorthand way of expressing the value of a set of skills that are valuable to a particular society at a particular time.

Education does not fill a bucket. It is one part of a multifactorial matrix within which we develop children's skills of thinking. Our goals in education are constrained not with reference to what is genetically possible, but with reference to what we are able at the present time to accomplish. We have avowedly enhancement-directed goals in parenting, and one of them is education. What differentiates the search for an IQ gene from the search for a good educational policy is the ability we have to predict and control the results of the activity, not the fundamental aim. And what frequently distinguishes the responsible parent from the irresponsible parent is the difference in the means and extent to which certain kinds of enhancement are pursued, not the difference between enhancement desires and "normal" parenting. Our inability to make the correct distinction here is one reason why it bothers us when parents shove law school down a child's throat, but it doesn't faze us a bit when they elect to use orthodontics or private schools.

If the distinction between interventions that are natural and those that are enhancement doesn't help us to solve problems, perhaps a distinction between therapy and enhancement can help. The idea is that while some interventions are therapeutic, and thus within the domain of medical practice, anything not properly thought of as curative or restorative belongs to a class we can call enhancement, and thus regulate in a special way. But how much work will this distinction really do? Much of family practice medicine is preventive in nature. It neither treats a disease nor restores a lost ability. Vaccinations and prenatal vitamins would count as enhancement on such a score, yet they seems an obvious part of appropriate medical practice. An even larger class of interventions responds to a deficit that may not yet cause the patient to suffer: is blood pressure medication an enhancement? Perhaps we should instead think of medicine as society's clinical attempt to keep patients within norms. Yet some therapies, such as reconstructive plastic surgery, respond to a deficit even where no clinical norm can be articulated. The development of norms, moreover, emerged not from some canon of disease and normalcy but from an evolving consensus among practitioners about what counts as a deficit.

Physicians have a tremendous amount of prestige in the community, and their careful, tentative pronouncements about disease and health carry a great deal of weight and get a lot of press. Many of the treatments designed by physicians are successful in alleviating symptoms in the ill patient. However, just as the past two centuries' movement toward "objective" diagnostic technologies encouraged the view that *disease* is prior to and more important than *illness*, disease and health are often assumed to be objective entities, waiting to be catalogued in a value-free book of pathologies.[14] The physician, we come to think, merely "discovers" a disease. This is true because, as John Dewey notes, the separation of experience and activities into distinct institutional compartments encourages the view that these divisions inhere *in experience itself.*[15] When medicine diagnoses an illness, or reports on standards of health, its results may be questioned, but we quickly come to think of the trait involved as an objective matter of "disease" or "health," forgetting that these categories are actually subject to our cultural evaluations of what kinds of traits are problematic or desirable in our lives.

The use of this hypostatized idea of the "medical" leads to the confused strategy invoked by Norman Daniels and others to distinguish enhancement from medicine with reference to a range of "species-typical functioning."[16] The idea of species-typical functioning is that from medical data and data "in society" we can find "a theoretical account of the design of the organism" that describes "the natural functional organization of a typical member of the species." We can then accord to medicine—and for Daniels this means "entitlement to access to medicine"—the task of bringing members of the species up this level, and to enhancement any attempt to exceed it. But this account assumes too much.

First, as just another version of the biological side of our "ontology of becoming," the species-typical account of normalcy posits without justification that members of the species are born with largely determined sets of capacities. Second, it depends on a strange idea of what counts as normal in medicine, exactly because it equivocates on the question of what medicine can tell us without reference to human social norms. According to this account, we first unearth data about what counts as an impairment, then round up everyone who is affected, and finally make sure that those with the most impairment receive the most treatment, while those who are unimpaired are entitled to no treatment. But who decides what to measure in the first place, or how to draw the line between impairment and enhancement in the raw data? As with most accounts that draw on a loosely Aristotelian biology, the species-typical functioning account assumes both that medicine can make empirical observations without reference to some set of norms, and that those observations can then be appropriated in a more or less equal way. The bottom line is that what is "species-typical" depends on how we measure and evaluate the species, which are normative endeavors.

In all of its incarnations, the second part of the ontology of becoming in our culture is the naive reliance on a hypostatized idea of what counts as medicine. We are not likely to deduce a clear meaning for medicine that derives from Hippocrates, God, or metaphysics. The fact is that our culture happens to choose to deal with certain kinds of problems in buildings called hospitals, where we locate people whom we have trained to do things that we see as desirable to our bodies. We experience some of those activities as curative and restorative, and the measure both of those activities and of their efficacy and desirability has been tested through a long history of research and scholarship yielding consensus on some issues. Activities that we do not experience as curative or restorative are not "bad medicine" except inasmuch as they are less effective than other available modes for dealing with the problems of our experience as embodied.

When new interventions are proposed, the correct question is not "Is this medicine or enhancement?" but rather "Will this approach to this issue work better than others?" Depression comes to mind. If we did not live in a world defined by and replete with mindless jobs, commodified sex, frozen pizza, and our friends at "Cheers," perhaps we hordes of the nicely neurotic would feel less alienated—even if many of us have a depression gene or are amenable to Prozac. The search for a depression gene seems to me the medical equivalent of the StairMaster, which allows us to gorge ourselves and live indolent daily lives in the faith that we can climb the "stairs to nowhere" to work it off. Our culture has a profound faith in quick fixes, and the success of some notable pharmacological "magic bullets" (aspirin, penicillin) has made this especially true for medicine of the past fifty years. Enhancements may be more or less appropriate on the basis of the wisdom of their use, and medical information can be helpful in determining such matters. However, the bald assertion that plastic surgery or designer drugs are "not medical" begs the question of how we choose to evaluate what counts as medicine in the first place. We get much more mileage out of judgments about interventions that depend on whether a particular method works well in clinical settings or is the most effective option in our social or personal quiver than judgments based on whether an intervention is properly classed as natural or medical.

Genetic Enhancement in Parenting

Parenthood sometimes feels like a laboratory in enhancement. All of us with children experience the pressure to develop the life of an infant, a young person, a young adult. Children present themselves to us as so many interwoven needs: for support, for care, for attention. The struggle to parent feels like a perilous and wonderful dance as we balance the need to transmit and inculcate values and culture with the need to give children what Joel Feinberg terms "an open future." As we make choices about our children, we pick up some cultural lessons that work not only for mundane parental decision making but also for the radical possibility of making, perhaps sooner than we think, some systematic choices about the enhancement of our children through genetic technologies.

It may turn out, in this quest for some social improvement, that genes are among the least effective tools for advancing personal, familial, and social goals. Technical failings in all previously initiated trials of gene therapy suggest that our powers to induce genetic modification have perhaps been exaggerated. Is it likely that even an effective genetic therapy would revamp the human species? Not especially. Nor is it likely, conversely, that altered genes will destroy our human natures. Conventional social institutions, such as schools and churches, have a much more immediate effect on who we become, and we "conventional" parents can botch up child-making quite well without gene therapy.

There is plenty, though, to be frightened about when conversation turns to eugenics. The fear is not of genetic control, it is of socially prescribed blueprints of perfection, enforced by intolerant scientists-cum-bureaucrats. We've seen the results in our own century, and can at least glean from the misadventures chronicled by Daniel Kevles and others that a scientifically styled "perfect society," stratified by genes, makes little sense in a world where genetic variability turns out to be a virtue—and in which specialization and rigidity spell extinction. There are also plenty of practical examples of the danger of replacing parental responsibility with overarching social control.[17]

How then can we put history's lessons to work in making responsible use of our social aim of improvement? First, we have to separate the dreams of eugenics from the hopes of families. The quest to improve humanity is not mere aberration, the deluded dream of social engineers. The *Newsweek* description of perfection (tall, blond, powerful, smart children, made-to-order) is shocking, in part, because it is lifted directly from fashion magazines and television. Our culture pursues notions of "perfection," from eye color to weight to "swagger." We invest billions of dollars in the attempt to make people more intelligent and less aggressive. We call this attempt public education. As with eugenics, the goal of education is to design and inculcate skills and norms in the behaviors of offspring, from sexual mores through beliefs about history to respect for the law. Athletic activity and school lunches are designed so that children will grow up to be stronger, more capable, and smarter. Those who do not perform well in school are "failed," and miss out on college, better-paying work, or social success. That families and the social order should abandon the aim at the improvement of children is unthinkable. Libraries, nutritional and environmental regulations, and the matrix of social and political institutions we have crafted testify to the necessity of this goal.

Because we make big social blunders, our programs, visionary plans, and political ambitions often do not provide the New World Order that is promised. Great plans for our children's futures can also be doomed by shortsightedness, avarice, and cowardice, or merely turn out to be unworkable or inapplicable to environmental and cultural conditions. Nonetheless, the hope for continuing improvement, "making the world better for our children," remains central to human progress and is present in the rhetoric of markets, politics, religion, and even medicine. We learn from our mistakes and work for a better future. Thus the deadly and not-so-deadly sins we need to avoid along the road to enhancement are not all related to genes, test tubes, or the Nazis. The five we explore here are instead sins we learn to avoid as parents and social stewards: calculativeness, overbearingness, shortsightedness, hasty judgment, and pessimism.

The Sin of Calculativeness

Consider, for a moment, your memories of childhood. Parents (or guardians) send children thousands of messages about appropriate behavior, communicating their hopes and fears. Some give an inordinate amount of advice and counseling.[18] Some even set up elaborate systems of rules and procedures to instill certain habits and values. You might have been awarded $5 for mowing the lawn, cleaning your room, and washing the dishes. You might have lost your driving or entertaining privileges for misbehaving. These thoughtful, organized systems provide a network of beliefs; they structure the developmental environment. But they are not the whole experience of being a child. In fact, you may have learned much more from the character, rules, and goals of your parents by watching what *they in fact did* than by obeying or disobeying the rules they set for you. Conversely, the most vital and formative experiences of your childhood may not have had anything at all to do with the detailed plans your parents agonized over. A brief, unpredictable outburst from a parent may outweigh years of regimented education. The sudden death of a grandparent or parent may change the entire family ethos. We commit the sin of calculativeness when we overemphasize the importance of planning and systematic choices in parenthood.

Like most sins, calculativeness is as much impractical as it is immoral. It is extraordinarily difficult to know what actions and words will register in the minds of our children. How will the whole package fit together: the way we treat them, the food we feed them, the genes we give them, and the rules we set for them? The most complex and sophisticated plans for a child's future can turn out to be the least effective, and we may send messages that are much more mixed than we know.

At times, we cannot even be sure what we want for (and from) our children. Children can be instruments in our own efforts to work out our childhood insecurities, ambitions, and fears. Our own frustrated effort to get to Harvard may become our child's yoke. Abuse by a father becomes a son's abuse of his own child. The approval of friends and neighbors can influence the way we dress and teach our children. Parents can effortlessly create tortuous paradigms that children are expected to meet.

Our beliefs about the "perfection" of our expected child may be much more simple or grand than we can articulate. The hopeful, infertile couple who expresses the fervent wish for any biological child, saying "all we want is a healthy baby," may not be fully conscious of the reasons why they seek not only health but also *biological* relation. A father who spends weeks teaching baseball to his son might actually prefer (at some deeper and more inarticulate level) that he and his son be able to have a nice conversation or share a common goal. Because parenting is subtle, sophisticated, and enormously complicated, it is not at all surprising that we should be unaware of our own motivations—or even that we should act in ways contrary to our deeply held desires. Parenting habits are as complex as any human patterns of behavior, and can be malleable or rigid, conscientious or the thoughtless repetition of our own parents' behaviors.

Though genetic tests and therapies may not have the capacity to advance the intelligence and attractiveness of our children, faith in the efficacy of genetic technologies could lead parents to deemphasize important parts of parental responsibility. In addition, a faith in genetic modifications of offspring could encourage the emphasis by parents on narrow, artificially defined traits. Parents could have hopes of transmitting, in a simple and systematic way, all of the currently fashionable traits to their children, relying on the common images of "perfection" in the public. These images of perfection are not taken from the dreams of dictators or science fiction novels. They are present in advertisements, polls, television programs, and movies.[19] The perfect baby of *Cosmopolitan* or *Men's Health* might grow to be 6 feet tall, 185 pounds, and disease-free. His IQ is 150, with special aptitudes in biomedical science. He has blond hair, blue eyes. He is aggressive and has superior ability in football, hockey, and basketball, but also enjoys poetry and fine wine.

The parent who opts for such systematic control over the creation of a child puts faith in the ability of "genetic parenthood" to create a child that has particular traits. The more ordinary ways of parenting offer no such systematic options. The hereditary possibilities in "conventional" parenthood revolve around a mixture of similarities (traits already in our family), over which we have little control. Will she have my ears or her mother's? Whose toes? We don't know, and we have no control over the answer. By contrast, genetic parenthood *seems* to offer a different kind of control. Here, parents could utterly abandon similarities, replacing them with choices that are *reasoned* in advance. If we thought we *could* systematically impart an IQ of 150 instead of whatever mental traits we carry, we might opt to change our hereditary gift.

The "sin" of these calculated choices is not rooted in the idea that they might actually work, giving our kids high IQs or the appearances of gods. That much is unlikely to emerge from the polymerase chain reaction (PCR), gene-splicing, and vector technology of 1995, and may be conceptually impossible for reasons we described above. The sin is in understanding a child to be the result of systematic choices, and thus allowing genetic choices to define the child's telos. The faith that genetic enhancements can alter character (removing homosexuality or increasing thoughtfulness) lends itself to a parenthood of oppressive control. Parents that choose traits as calculative consumers might come to devalue the essential connections of relatedness and sameness in the family relationship.

Though it may not be articulated in the fashion of the day, parents also want their children to be like themselves. This is evidenced by the celebration of every child as a "perfect" child, beautiful and appropriate exactly because it represents the particular union of two particular people. We share names and houses and values with our children, as well as important biological and cultural habits. The essential fact of this sharing is not its biological element. Adoptive parents also appreciate similarities in their children, and secure it through familial patterns of value-transmission. The sharing of similarities among members of a family could be diluted by genetic choices. A parent who is expecting a "brilliant child" could value that child only for her accomplishments, rather than for her struggles and growth.

And there is the problem of efficacy in our calculations: whatever our social goals, the likelihood of achieving them through genetic interventions does—or should—figure in calculations about how to spend money. This propensity should then be measured against other means of dealing with the problem. In the case of intelligence, it is amazing that we are willing to spend millions of dollars on the search for genes that code for calculative efficiency while Head Start programs go unfunded, teachers are underpaid and overworked, and even smart kids graduate ill prepared for the job market and uninspired by democracy.

The Sin of Being Overbearing

Hans Jonas and Joel Feinberg refer to a child's right to be open to as much freedom of identity as possible. They fear that genetic engineering, by stylizing children along the lines of rigid parental expectations, could steal this right. Children would be born into a world where their ultimate choices have been made by parents before the moment of their birth. While Jonas's fear hinges, in part, on the power of genetics to accomplish this feat, his insistence on children's continuing need for freedom is important. Genetic expectations, we noted above, could carry tremendous weight, as parents hope that children will become the sort of person whom they engineered. Already, parents who use IVF technologies to implant the sperm of especially intelligent or athletic donors have expressed expectations of greatness from such children, insisting on endless piano lessons or daily tennis practice.

How do we distinguish between responsible hopes and overbearing ambitions in reproductive enhancement? A pragmatic answer begins with the recognition of essential continuity between hopes connected to genetic engineering and everyday hopes. The parent who wants a beautiful ballerina will want one whether or not genetic technologies are in the picture. Likewise, parents whose guiding motivation is that a child find and pursue some kind of flourishing will be reluctant to use genetic improvements or conventional means of overdetermining identity through reproduction. The decisions of parenthood are not always explicit, and take place in a social context, so that parents are constantly exposed to suggestions from all quarters about the kind of baby that is "good." Fortunately, there are also extensive pressures in society that push for the maintenance of randomness and the celebration of hereditary difference. The sentiment that each baby is "perfect" conveys this pressure, as does the choice many parents make in refusing unnecessary ultrasound exams, waiting until birth to know a baby's gender.

We have to emphasize the responsibility that comes with new information before we spill it onto the table and write it into the chart. Parents must ask themselves of each new test and procedure, "why do I want to know X?" Honest answers may turn up more than parental curiosity. If tests for gender, intelligence, and other traits cultivate a parental mentality in which traits take center stage, it pays to consider the danger of such planning and expectations. However, as John Stuart Mill, William James, and more recently Derek Parfit have made so plain, there must be tolerance to different ways of approaching human natures. Wherever tests and procedures do not compromise the child, plural approaches to genetic modifications must be allowed. No simple, single solution will work. It makes no sense, and is generally counterproductive, to issue wholesale policy restrictions of any genetic research that is "positive" or "enhancing" in character.

Experiments in biological engineering must be tempered by the respect for diversity and for each individual child. Overbearing parents can reduce the child to an instrument of their own ambitions or insecurities. This is no more appropriate when exercised through genetic technologies than when implemented by a parent who insists that a child accompany him to Klan meetings or who refuses appropriate medical treatment in the name of religious beliefs a child cannot endorse. Children must be allowed to imagine and grow, and the balance to be struck is between instilling the values that parents hold and allowing the growth that could pull children away from those values. The desire for sameness can be a crippling expression of parental ego, just as the desire for a fashionably beautiful child can express self-loathing in the parent. The key is to avoid extreme measures through biological *or any other means*, and to temper decisions before birth with the recognition that every child has a right to make some decisions about her own identity.

The Sin of Shortsightedness

As much as we plan for and anticipate the future, we cannot be sure what our children should or will become. We simply cannot anticipate the world of tomorrow. Within the past decade, an empire has been destroyed, Europe has formed an economic alliance, genetic testing has been developed, and computer speed has increased 10,000-fold. Economic and political prophets failed to predict a major market crash, the United States went to war with a Third World country, and a U.S. physician began an assisted suicide delivery service. Fashions have changed, as have language, science, philosophy, psychology, and secondary education. Our heroes have also changed: Alan Alda was in, then out; George Bush's 80 percent approval rating dropped to 34 percent in less than a year. What will the next decade, a mere ten years in the life of a child, hold in store? If you think you know, odds are you have a shortsightedness problem. Which is fine—unless it becomes the basis for designing your descendants.

One advantage of the conventional uncertainties in parenting is that just about all our rules and practices can be changed to fit the exigencies of a changing world. For example, business schools grew to their apex during the early 1980s, then began to shrink as fewer employers recruited business majors. Savvy students quickly transferred from "entrepreneurship" into the humanities and environmental sciences. Parents with stubborn, outmoded commitments to business school for their kids ended up with unemployed progeny or children on Prozac. Younger children are even more malleable than college students, and infants will accept the most conditioning of all. A child is receptive to language, math, rules, values, and abstract ideas. If conditions change, a child adapts. One danger of genetic engineering for positive traits, then, is the sin of shortsightedness: how can we know which traits to lock-in through genetics in a world where fashions fade quickly and rigidity is a disadvantage?

An intelligent approach will militate against hasty and acontextual decisions. Just as it is difficult to plan the inculcation of values and character in children, hard to know what action or word will register, so too is it difficult to single out characteristics that will make a child's life better. In H.J. Müeller's *Out of the Night* (1935), one of the most important eugenic treatises, the geneticist argued in favor of breeding children who embodied the traits of Vladimir Lenin and Karl Marx (among others). When these and his theory failed to curry favor with Stalin, he dropped references in his lectures to Lenin and Marx for more fashionable figures. Political currency plays a role in our notions of perfection.

When we examine contemporary genetic optimists' plans for a gradual but total revision of human natures, what is most striking is their confidence that we already have the wisdom to select the best traits. Like Plato, writers such as Leroy Hood and Brian Stableford[20] assume that human natures are immutable and determined prior to birth, so that genetic engineers have merely to figure out how to manipulate stable biological materials in order to accomplish wondrous things. A human with scales and gills would help us to live in the sea, Stableford writes, where we would be able to exploit its unending resources. But to which oceans does Stableford refer? We have turned much of the sea into a colossal dump for industrial and commercial waste. How much would we have to give up to live in this deep, dark ocean? Why would we want to live there? The description of genetic engineering as a one-stop shop for human improvement sometimes depends on wildly unrealistic political and scientific plans. Such grand schemes are not only difficult from a genetic standpoint, they can simply be icons of poorly thought-through political visions for human growth.

A parent who desires a smart child might actually be able, at some point, to increase the calculative speed of that child's brain. At present, scientists often compare the power of our minds to the power of computers. Computers are better when they are faster, so much of this research has focused on a faster brain. In ten years, though, it may turn out that calculative speed is a hindrance to thoughtfulness, imagination, and vision. A child could thus be robbed of the ability to adapt, stuck with a trait that hinders her ability to work with flexibility in the changing world. All this while expected by her parents to be brilliant.

Moreover, it is not always wise to assume that "more of a good thing is better." Genetic diversity has tremendous value because it provides the opportunity for those of many hereditary backgrounds to employ differing approaches toward maximization of the potential of a given environment. If dozens of children were created from the genes of an Einstein, would the world be a better place? Albert Einstein was the product of a particular set of parents, experiences, and inspirations. Growing up in suburban Dallas as the child of an oil baron, a cloned Einstein might as easily end up driving a truck or selling horizontal drilling rigs. He might live alone and homeless. Even with an optimal environment, young Clonestein would find that his progenitor's approach has been all but replaced by a different mode of analysis, as differing approaches to problems rendered his style of physics less capable of explanation and control.

Just as it is important for parents to allow children to develop in individual ways, there is reason for parental plans to allow for a changing world. Highly directed parental ambitions for children, such as success in a particular sport or with a particular musical instrument, can result in crushed hopes for parent and child. There are only so many slots on college and professional basketball teams, and not many will go to five-foot seven-inch men. Only one in a million musicians attends Julliard. It would be no advantage to choose male offspring, which most Americans report that they would, if suddenly 60 percent of live births were male.[21]

Children need support—not pressure—in pursuing their own dreams within the context of family and culture. Diffuse parental hopes are more appropriate. Children need to learn courage and self-esteem, and need to be critical and functionally literate. They should have the support of their parents as they learn and grow.

The Sin of Hasty Judgment

In College Station, Texas, there are acres and acres of "test fields." In these fields, students at Texas A&M see to it that there are more hybridized and genetically engineered crops than in any other region on earth. It is here that the super-tomato was born. Cantaloupes are genetically crossed with watermelons, and cows have been cloned and genetically modified in literally thousands of ways. College Station is also home to amazing new strains of disease, which began to thrive on these same new crops. Genetic engineering in agriculture has been a proving ground for the possibilities of modification of humans. The results are somewhat revealing: genetically engineered fruits and vegetables are frequently much more vulnerable to diseases and parasites, and rarely taste as good as nonhybridized, nonengineered strains.[22] Even when these strains work out, the monolithic use of single hybrids reduces diversity, encouraging new pests and reenergizing long-lost plagues. Engineering of plants and animals can also result in the transmission of dangerous materials into the human and animal food supply.

The perfect baby, like perfect soybeans and perfect corn, could turn out to be markedly imperfect. How difficult would it be to live an engineered life? Hans Jonas cautions of the danger of freakish accidents in genetic engineering, of the kind discussed in the 1970s controversy over genetic engineering played out in citizens' hearings in Cambridge, Massachusetts, and the temporary self-imposed moratorium declared by scientists after meeting at Asilomar.[23] Ironically, the more important accidents may be more likely to occur *after* the birth of an apparently healthy, improved baby. While medical technologies could make alterations in the physical characteristics of a newborn, we can hardly hope for the viability of those traits in our complex world. For example, wild strawberries have a much better chance of surviving against infection and parasites than engineered strawberries do. The reason is that while genetic engineers controlled for particular traits, they could not control for the dozens of conditions that a strawberry faces. Wild strawberries pack a variety of genetic habits. These "resistances" help them to have stable interaction with a range of circumstances. A genetically engineered strawberry, on the other hand, is a hit-or-miss proposition, with engineering emphasis placed only on particular traits.

A child who is engineered to possess positive traits might end up suffering unexpected and disastrous ills. It is extremely dangerous to move too quickly in the direction of changing human traits, lest we forget to control, or forget that we can't control, for the vast variety of human environmental conditions. Just as the gene that presumably causes sickle cell anemia codes for resistance to malaria, the gene for sonar hearing might interfere with the genetic pattern that codes for opposable thumbs or sex organs. In a strawberry, such mistakes can lead to new diseases and bad-tasting fruit; in a human child, such errors become the sin of hasty judgment, and could be much more catastrophic for families.

There is also a more general point to be gleaned from our recent experience with agricultural engineering. The genetically enhanced tomato was delicious and tender when raised in lab conditions, but turned out to taste rubbery in real life. Seedless watermelons also seem to suffer from diminished flavor. By analogy, imagine the beautiful, intelligent, even-tempered girl developed by genetic engineering. Could she survive in an imperfect world, with bad water and fatty foods? Would others hate her or envy her? On paper, genetic engineering's traits look enticing. In practice, the attractiveness of other people is more random and depends on their quirks as well as assets. A perfect child could find the world of imperfection, disease, disasters, and emotions deadly or unsatisfying.

A pragmatic approach urges more cautious progress toward improving humans. Just as parents should promote malleability in their parenting, there must be room for imperfections and developmental choices. The child who is genetically crafted to 1996's models of perfection may find the world of 2014 intolerable. Instead, parents should aim to continue to update their style of parenting to match the demands of natural and social conditions.

This means that some modifications may indeed become advisable, but only on condition of reversibility. It might be to our advantage to have access to Rostand's "built-in cheek headlights" at some point during our lives. However, we would want to insist on the reversibility of the modification, and to carefully examine its side effects prior to clinical trials.

The Sin of Pessimism

In his essay "The Moral Equivalent of War," William James argues that while war is to be avoided at all costs, humans seem to need to exercise aggression and domination during their lives. He termed the channeling of these powerful impulses into other activities "the moral equivalent" of war. This notion of moral equivalency is useful here. Reproductive genetic enhancement may present new choices, but these choices are suffused with the moral equivalence of activities already present in the context of parenthood. The moral dominion of parenthood creates the context for reproductive genetic interventions. Thus while caution is intelligent, we need not treat genetics as a radically different endeavor, a slippery slope to biological castes and Frankenstein. The categorical opponents of genetic enhancement, Paul Ramsey and Jeremy Rifkin being the most notable, have utilized rigid rules to enforce the sanctity of human genetic coding. Such an ethic does little to guide our actions—it is simply naive in the light of other social pressures to apply scientific results, obtain improvements in life, and have healthy children. Ethics cannot ignore science: the problem with putting the values that are present in our culture to *use* in our culture is that those values can sometimes be "undermined by the conclusions of modern science."[24] Values that are rooted in the ontology of becoming are useless simply because that ontology is rooted in a shallow understanding of science.

We also do well to consider John Dewey's charge that "if intelligent method is lacking, prejudice, the pressure of immediate circumstance, self-interest and class interest, traditional customs, institutions of accidental historical origin, are *not* lacking, and they tend to take the place of intelligence."[25] For example, the few fetal diagnoses available now are so expensive that only the wealthy use them. As a consequence, a disproportionate number of children with Down's syndrome are "almost certainly born to the less affluent."[26] Our claim that some eugenic selection is already present in social engineering intimates another danger, then, of uncritical genetic research: it may be engineering that benefits only the powerful and wealthy. If society chooses not to concern itself with reproductive enhancement, we too have made a choice: to leave science to the scientists, and its application to political pressure and happenstance. Consider the application of genetic research in its political and economic context: where there are therapies, there will always be pressures on a physician to offer them. The day that a gene for homosexuality is announced is too late for bioethics to put a "spin" on whether that gene is useful. We need to join the conversation about appropriate research before it becomes technology.

If pessimism is sinful, though, we have also seen that abject optimism is not its antidote. Even assuming that certain isolable ailments could be dealt with by genetic engineering, the approach to avoiding the not-so-deadly sins must be intelligent and cautious; we work toward developing protocols and therapies *experimentally* and *gradually.* This approach takes seriously the caution implicit in the hands-off attitude of those who would leave genetics to nature without surrendering the hope to make our condition and our nature better a little at a time. Social conversation concerning the enhancement of children is possible, and technological advancement is desirable in pediatrics. First, though, bioethicists' conversation about expensive and sophisticated genetic technologies must be connected to public conversations about parenthood. This requires us to abandon the search for an exotic ethics of enhancement, and get our hands dirty in the mundane world of the ordinary parents who will make decisions about genetic enhancement.

Notes

Chapter 1 *The Landscape of Genetic Technology*

1. Michael R. Cummings, *Human Heredity: Principles and Issues* (St. Paul: West, 1988), 3. Cf. A. Corcos, "Reproductive Hereditary Beliefs of the Hindus, Based on Their Sacred Books," *Journal of Heredity* 75 (1984): 152–54.

2. Kenneth Korey, *The Essential Darwin* (Boston: Little, Brown, 1984), xvii.

3. As becomes clear later, John Dewey understood this advance to be central to the success of future philosophy: the naturalization of the notion of species was, for Dewey, a step toward acknowledging the relationship between experience and nature, heredity and morals. But Dewey did not celebrate the emphasis in later Darwinian thinking on a determinate structure of competition. Where later Darwinism emphasized the cutthroat power of kill-or-be-killed competitiveness, Dewey emphasized the value and power of cooperation and of democracy.

4. Though Darwin did not write as a social theorist, he was deeply influenced on this point by Thomas Malthus and the Scottish economists Adam Smith and Thomas Hutchinson.

5. Richard C. Lewontin et al., *Not in Our Genes: Biology, Ideology, and Human Nature* (New York: Random House, 1984), 57.

6. Daniel J. Kevles, *In the Name of Eugenics: Genetics and the Uses of Human Heredity* (Los Angeles: University of California Press, 1985), 4.

7. Kevles, *In the Name of Eugenics*, 116.

8. Kevles, *In the Name of Eugenics*, 117.

9. Kevles, *In the Name of Eugenics*, 117.

10. We will treat this issue in detail in Chapter 6. See Dorothy Nelkin, *Dangerous Diagnostics: The Social Power of Biological Information* (New York: Basic Books, 1989), esp. 11–24, 164–70.

11. Jean Bethke Elshtain, *The Family in Political Thought* (Amherst: University of Massachusetts Press, 1982), 288.

12. Louise Erdrich, "A Woman's Work," *Harper's Magazine* (May 1993): 35.

13. U.S. Department of Energy, *Human Genome 1991–92 Program Report*, 1992.

14. *Human Genome 1991–92*, 3–5.

15. See his own account of the social and scientific pursuit of the double helix in James Watson, *The Double Helix* (New York: Norton, 1980). For commentary on the role of Watson in the genome project see Robert Wright, "Achilles' Helix," *New Republic* 21 (July 9, 1990): 21–31.

16. This is only a gloss of a much more extensive and complex argument. The clearest exposition is found in R.C. Lewontin et al., *Not in Our Genes*. There is a concomitant problem, as well, with the assumption of a stable relationship between

genotype and phenotype: not all organisms with the same genotype have the same phenotypic expression. We discuss this problem in detail in Chapter 5.

17. "Gene Therapy's Leap Beyond the Lab," *U.S. News and World Report* 82 (May 10, 1993): 82.

18. "Gene Therapy's Leap," 82.

Chapter 2 *The Magic Answer? Hopes for Genetic Cures*

1. Thomas Lee, *The Human Genome Project: Cracking the Genetic Code of Life* (New York: Plenum, 1991), 1.

2. Dean Hamer, et al., "A Linkage Between DNA Markers on the X-Chromosome and Male Sexual Orientation," *Science* 261 (1993): 321–27; S. Hu, et al., "Linkage Between Sexual Orientation and Chromosome Xq28 in Males But Not in Females,"*Nature Genetics* 11 (1995): 248–56; Andrea Pattatucci and Dean Hamer, "Development and Familiality of Sexual Orientation in Females," *Behavior Genetics* 25 (1995):407–20.

3. Brian Stableford, *Future Man* (New York: Crown, 1984), 13–15.

4. Stableford, *Future Man,* 14–15.

5. Stableford, *Future Man,* 108.

6. Stableford does not suggest, as has recently been suggested by some Atlanta groups, that we should begin to favor darker skin to cope with the depleting ozone layer and the dangers that holds for Caucasians.

7. Stableford, *Future Man,* 119–23.

8. See Arthur R. Jensen, "The Current Status of the IQ Controversy," *Australian Psychologist* 13 (1978): 7–27. Also Robert A. Wallace, *The Genesis Factor* (New York: Morrow, 1979), 12–15.

9. Jean Rostand, *Can Man Be Modified?: Predictions of Our Biological Future,* translated by Jonathan Griffin (New York: Basic Books, 1959), 57–58. Of course such dreams, as we detailed in Chapter 1, are much older than Nietzsche. Rostand is pointing to the renaissance of descriptions of "super persons" and "last men" that occurs in the nineteenth century.

10. See not only Richard Lewontin's analysis in *Biology as Ideology: The Doctrine of DNA* (New York: Harper Perennial and Harper Torchbooks, 1991), 67, but also the Department of Energy*Human Genome 1991–92 Program Report,* 1992, 218. This issue also links the question of genomic "information" to the issue of "virtual" reality: is a clone a "virtual" person, or a real one? The so-called information "superhighway" also makes profound promises along the lines of the genome project. We are promised a better world through additional mediation: if we can encode our messages in data, transmit them through four or five different computer links and satellites, then translate them back into images, we will all be "connected" on a massive new highway. The central part of this claim, of course, is that *data* or *information* is the key to immediacy with friends and neighbors around the world. Similarly, the genome project's "genetic information" promises a genetic "editing machine" that is the key to managing reproductive affairs in a new way.

11. See Shulamith Firestone, *Dialectic of Sex* (New York: Morrow, 1970).

12. Robert Nozick, *Anarchy, State and Utopia* (New York: Random House,

1974), 315. For a detailed analysis, see Jonathan Glover, *What Sort of People Should There Be?* (England: Penguin Books, 1984).

13. Kenneth Ryan, "Ethics and Pragmatism in Scientific Affairs," *BioScience* 29 (1979): 35–37.

14. One example might be the awe-inspiring three-ring circus assembled for the Human Genome Project-funded PBS special, "The Secret of Life."

15. Ryan, "Ethics and Pragmatism," 36.

16. Ryan, "Ethics and Pragmatism," 37.

17. Ryan, "Ethics and Pragmatism," 37.

Chapter 3 *Playing God? Fears about Genetic Engineering*

1. Cited in Phillip Elmer-Dewitt, "The Genetic Revolution," *Time* 143 (January 17, 1994): 46.

2. Jeremy Rifkin, *Algeny: A New Word—A New World* (New York: Penguin Books, 1983),65.

3. Rifkin, *Algeny*, 71.

4. Rifkin, *Algeny*, 72.

5. Rifkin, *Algeny*, 159.

6. Rifkin, *Algeny*, 160.

7. Rifkin, *Algeny*, 208. On the history of biological systems theory, see E.J. Dijksterhuis, *The Mechanization of the World Picture: Pythagoras to Newton* (Princeton: Princeton University Press, 1976).

8. Rifkin, *Algeny*, 227.

9. Rifkin, *Algeny*, 230.

10. Rifkin, *Algeny*, 231.

11. Rifkin, *Algeny*, 232.

12. Rifkin, *Algeny*, 251.

13. Rifkin, *Algeny*, 252.

14. Robyn Rowland, *Living Laboratories: Women and Reproductive Technologies* (Bloomington: Indiana University Press, 1992), 81. See also Gwynne Basen et al., *Misconceptions: The Social Construction of Choice and the New Reproductive Technologies* (Quebec: Voyageur, 1994).

15. Rowland, *Living Laboratories*, 13.

16. Rowland, *Living Laboratories*, 14.

17. "Infant Sex Pre-selection Controversy Aids Marketeers," *Health Care Marketing Report* 89 (1986): 7.

18. Rowland, *Living Laboratories*, 82. Rowland claims an 85 to 95 percent success rate for this protocol.

19. Robyn Rowland, "Motherhood, Patriarchal Power, Alienation, and the Issue of Choice in Sex Preselection," in *Man-Made Women: How New Reproductive Technologies Affect Women* (Bloomington: Indiana University Press, 1987), 81.

20. Rowland, "Motherhood," 83.

21. Quoted in Rowland, *Living Laboratories*, 90.

22. Rowland, *Living Laboratories*, 90.

23. Approximately 100 of the Fortune 500 companies engage in preventive

screening. See Jane McLoughlin, "Playing Russian Roulette with Employees," *Guardian* (July 10, 1986): 10.

24. Rowland, *Living Laboratories*, 112.

25. Hans Jonas, "Biological Engineering—A Preview," from his *Philosophical Essays from Ancient Creed to Technological Man* (Englewood Cliffs, N.J.: Prentice-Hall, 1974), 140–41.

26. Jonas, *Philosophical Essays*, 143.

27. Jonas, *Philosophical Essays*, 144.

28. Jonas, *Philosophical Essays*, 145.

29. Jonas, *Philosophical Essays*, 152.

30. Jonas, *Philosophical Essays*, 153.

31. Jonas, *Philosophical Essays*, 160–61.

32. Jonas, *Philosophical Essays*, 166.

33. Paul Ramsey, *Fabricated Man: The Ethics of Genetic Control* (New York: Yale University Press, 1970), 138, 143.

34. This "three-implications" analysis is adapted from Stephen P. Stich, "The Genetic Adventure," *Report from the Center for Philosophy & Public Policy* (1982): 10–11.

35. Cf. C. Keith Boone, "Bad Axioms in Genetic Engineering," *Hastings Center Report* (August 1988): 10.

36. Paul Ramsey, "Shall We 'Reproduce'?" *Journal of the American Medical Association* 220 (June 12, 1972): 1484.

37. Leon R. Kass, *Toward a More Natural Science: Biology and Human Affairs* (New York: Free Press, 1985), 25.

38. Kass, *Toward a More Natural Science*, 26.

39. Kass, *Toward a More Natural Science*, 98.

40. Kass, *Toward a More Natural Science*, 25.

Chapter 4 *Debunking the Myths*

1. Efforts to replace this method with the study of ethical reasoning and detailed study of cases have met with some success in genetics, as reported by L. S. Parker, "Bioethics for Human Geneticists: Models for Reasoning and Methods for Teaching," *American Journal of Human Genetics* 54: 137–47. Nonetheless, most of responsible research training remains unsuccessfully paternalistic. See G. McGee, "A Bill of Rights for Student Scientists," *Chronicle of Higher Education*, August 2, 1996.

2. John Dewey, *Logic: The Theory of Inquiry* (Carbondale: Southern Illinois University Press, 1991), 32.

3. John Lachs, *Intermediate Man* (Indianapolis: Hackett Press, 1981), 27.

4. Lachs, *Intermediate Man,* 57.

5. John Dewey, *Human Nature and Conduct* (New York: Holt, 1930), 10.

6. See Daniel B. McGee's analysis of the differences between the posture of making and having children: "Making or Having Babies," in William H. Brackney, ed., *Faith, Life, and Witness: The Papers of the Study and Research Division of the Baptist World Alliance* (Birmingham: Samford University Press, 1990), pp. 179–87. McGee claims, correctly, that the posture of having, while recognizing the fundamental

"moral luck" implicit in parenthood, takes inadequate account of the constant and appropriate "making" of children through parental activities.

7. Shulamith Firestone, *Dialectic of Sex* (New York: Morrow, 1970), 271.

8. Robyn Rowland, *Living Laboratories: Women and Reproductive Technologies* (Bloomington: Indiana University Press, 1992), 5.

9. Richard C. Lewontin, *Biology as Ideology: The Doctrine of DNA* (New York: Harper Perennial and Harper Torchbooks, 1991), 23.

10. Richard Lewontin, *Human Diversity* (New York: Scientific American Library, 1982), 19.

11. Lewontin, *Biology as Ideology.*

12. Lewontin, *Biology as Ideology.*

13. Lewontin, *Biology as Ideology.*

14. Lewontin, *Biology as Ideology.*

15. An even deeper study of the presence and prevalence of this metaphor is found in E. J. Dijksterhuis, *The Mechanization of the World Picture: Pythagoras to Newton* (Princeton: Princeton University Press, 1976). On its social impact, see John Dewey and J. H. Tufts, *Ethics* (New York: Holt, 1932), 353.

16. In medicine, economics, philosophy, ethics, and science it has always been so. It is extraordinarily difficult to free ourselves from the metaphors most readily available. If I spend my time writing and working on a computer, how surprising is it that I should use the language suggested to me by that new technology?

Chapter 5 *Biology, Culture, and Methodical Social Change: A Pragmatic Approach to Genetics*

1. John Dewey, *Logic: The Theory of Inquiry* (Carbondale: Southern Illinois University Press, 1991), 34.

2. Genetic engineering via food intake could turn out to be as promising as, if less exotic than, laboratory modification of the germ plasm.

3. Dewey, *Logic*, 49.

4. Dewey, *Logic*, 48.

5. Dewey, *Logic*, 49.

6. See R. W. Sleeper, *The Necessity of Pragmatism: John Dewey's Conception of Philosophy* (New Haven: Yale University Press, 1986), 142–44.

7. Adolf Portmann also makes this claim in *A Zoologist Looks at Humankind*, trans. Judith Schaefer (New York: Columbia University Press, 1990), 149.

8. See Sandra Rosenthal on this point, in *Speculative Pragmatism* (Amherst: University of Massachusetts Press, 1986), 16–18.

9. For an account of the workings of this theory in the thought of C. S. Peirce, see John E. Smith, *Purpose and Thought: The Meaning of Pragmatism* (New Haven: Yale University Press, 1978), esp. 53, 17.

10. Richard Lewontin, *Biology as Ideology: The Doctrine of DNA* (New York: Harper Perennial and Harper Torchbooks, 1991), 51–52.

11. James Campbell, *The Community Reconstructs: The Meaning of Pragmatic Social Thought* (Urbana-Champaign: University of Illinois Press, 1992), 46.

12. Campbell, *The Community Reconstructs*, 45.

13. Dewey, *Logic*, 76.

14. Jean Bethke Elshtain, *Rebuilding the Nest: A New Commitment to the American Family* (Milwaukee: Family Service America, 1990), 103.

15. Daniel B. McGee, "Making or Having Babies," in William H. Brackney, ed., *Faith, Life, and Witness: The Papers of the Study and Research Division of the Baptist World Alliance* (Birmingham: Samford University Press, 1990), 181.

16. Stephen Garber et al., "The Parent Trap: Perfection and Goals," *Sky* (September 1993): 26–30.

Chapter 6 *Genetic Approaches to Family and Public Health*

1. Louise Erdrich, "A Woman's Work," *Harper's Magazine* (May 1993): 35. Erdrich's discussion of pregnancy, birth, and moral bearing in parenthood is excellent and informs my treatment of the issue.

2. See S. Kay Toombs, *The Meaning of Illness: A Phenomenological Account of the Different Perspectives of Physician and Patient* (Dordrecht: Kluwer, 1992), 90–98. Also, for a pragmatic account of this phenomenon, see John J. McDermott, "Experience Grows By Its Edges: A Phenomenology of Relations in an American Philosophical Vein," in his *Streams of Experience: Reflections on the History and Philosophy of American Culture* (Amherst: University of Massachusetts Press, 1986), 141–56. In that piece, McDermott also cites Richard Zaner, *The Context of Self: A Phenomenological Inquiry Using Medicine as a Clue* (Athens: Ohio University Press, 1981), in an effort to connect a phenomenological approach to illness to a pragmatic approach to medicine.

3. William F. Bynum, *Science and the Practice of Medicine in the Nineteenth Century* (New York: Cambridge University Press, 1994).

4. Quoted in Mark Sullivan, "In What Sense Is Contemporary Medicine Dualistic?" *Culture, Medicine, and Psychiatry* 15 (1986): 10.

5. H. Tristram Engelhardt Jr., "Ideology and Etiology," *Journal of Medicine and Philosophy* 5 (1980): 256–63. Richard Zaner and Marx Wartofsky suggest, as an alternative, that the real shift is a replacement of the sort of inquiry conducted in the human sciences by the sort of inquiry conducted in the natural sciences. See their "Editorial: Understanding and Explanation in Medicine," *Journal of Medicine and Philosophy* 5 (1980): 2–3.

6. Toombs, *The Meaning of Illness*, 76.

7. Victor McKusick, *Mendelian Inheritance in Man* (Baltimore: Johns Hopkins University Press, 1983).

8. Information for this section distilled from J. S. Thompson, *Genetics in Medicine* (New York: Saunders, 1980), and J. deGrouchy, *Clinical Atlas of Human Chromosomes* (New York: Wiley, 1984).

9. Also unintentional, of course, is the social by-product of this policy: uninsured patients "drop in" to emergency rooms with their complex, early births. The resulting neonatal care costs dozens of times more than prenatal care, but is (in most states) the only free care available to the uninsured pregnant mother.

10. Cf. Tom Murray, "Genetics and the Moral Mission of Health Insurance," *Hastings Center Report* (December 1992): 12.

11. Nancy Kass, "Insurance for the Insurers: The Use of Genetic Tests," Hastings Center Report (December 1992): 6.

12. Kass, "Insurance for the Insurers," 8.

13. Neil Holtzman emphasizes that genetic tests are virtually unregulated, and that their commercial application is marred by unreliable instructions and bad manufacturing. See his "Research Discoveries and the Future of Screening," in B. M. Knoppers and C. M. Laberge, eds., *Genetic Screening: From Newborns to DNA Typing* (New York: Excerpta Medica, 1990), 354, 291–323.

14. In *Human Genome News* 5 (January 1994): 1–2, the results of ELSI studies on cystic fibrosis testing are reviewed. It is concluded that to test 3,500 individuals would cost approximately $400,000; in a test of 3,500, only *one* fetus with CF is likely to be identified.

15. Murray, "Genetics and the Moral Mission," 15.

16. See David Suzuki and Peter Knudtson, *Genethics* (Cambridge: Harvard University Press, 1990), 166.

17. A review of these market forces in biotechnology is useful. See Gregory Brown, "1994 Should See Firming Fundamentals," *Biotech* 12(1994): 20–21.

18. The tentative division of gene therapies into four categories is amalgamated here from Stephen Stich, "The Genetic Adventure," *Report from the Center for Philosophy and Public Policy* (1982):10–11, and from Daniel B. McGee, "Making and Having Babies," in William H. Brackney, ed., *Faith, Life, and Witness: The Papers of the Study and Research Division of the Baptist World Alliance* (Birmingham: Samford University Press, 1990), 179–81.

19. As described in Chapter 1.

20. Other examples include recent trials of interleukin-2 in the treatment of kidney and other forms of cancer; see "Interleukin-2," *FDA Consumer* (April 1994): 25–27. A general list of these materials is found in Mike Ginsburg, "Product Update," *Biotech* 12 (1994): 243–44, 358–59.

21. See Chapter 3, also Suzuki and Knudtson, *Genethics*, 183–84.

22. Suzuki and Knudtson, *Genethics*, 184.

23. This is perhaps a much more difficult issue than is at first apparent. As Stanford University philosopher Michael Bratman notes, if we determine that an aptitude to jazz music comes attached to the same gene that causes late-onset Alzheimer's, who is to say the music isn't worth it?

Chapter 7 *The Not-so-Deadly Sins of Genetic Enhancement*

1. Such a reference takes place in clinical language, such as the Apgar score assigned to infants at birth to quantify their likelihood of survival.

2. Alternatively, following S. Kay Toombs, illness becomes illness when it interrupts the routine ways in which we relate to our bodies in the world.

3. We specifically reviewed parts of this history as related to molecular and population genetics in Chapter 1, and examined medicine's evolving modes of perception and diagnosis in Chapters 2 and 5. In Chapter 6, we reviewed possible alternative approaches to diagnosis and medical training.

4. Recall our analysis of the evolving role of medical perception in Chapter 5.

5. Assuming, of course, that the components of patient experience that we discussed in Chapter 5 and Chapter 6 are taken into sufficient account.

6. William James, *The Principles of Psychology*, 3 vols., in *The Works of William James*, F. Burkhardt ed. (Cambridge: Harvard University Press, 1981).

7. Harry Stein, *Ethics (and Other Liabilities): Trying to Live Right in an Amoral World* (New York: St. Martin's, 1982), 167.

8. Note our reference in Chapter 4 to the general failure of chemical treatments for depression, for example.

9. Neil Holtzman describes this process in "How Technology Becomes Routine Procedure: The Case of DNA-Based Tests for Genetic Disorders," *Nucleic Acid Probes in the Diagnosis of Human Genetic Disease* (New York: Alan Liss, 1988).

10. Peter Kramer, *Listening to Prozac* (New York: HarperCollins, 1994).

11. Lawrence H. Diller, "The Run on Ritalin: Attention Deficit Disorder and Stimulant Treatment in the 1990s," *Hastings Center Report* 26, No. 2 (1996): 12.

12. Cf. Kathy Davis, *Reshaping the Female Body: The Dilemma of Cosmetic Surgery* (New York: Routledge, 1995).

13. Richard Lewontin, *Biology as Ideology: The Doctrine of DNA* (New York: Harper CollinsPerennial and Harper Torchbooks, 1991), 35.

14. Cf. Arthur Caplan, "Concepts of Health, Disease, and Illness," in the *Encyclopedia of the History of Medicine.*

15. John Dewey, *Art as Experience* (New York: Capricorn Books, 1958), 20–34.

16. Norman Daniels, *Just Health Care* (New York: Cambridge University Press, 1986), 28.

17. Cf. C. K. Chan, "Eugenics on the Rise: A Report from Singapore," in Ruth Chadwick, ed., *Ethics, Reproduction, and Genetic Control* (New York: Croom Helm, 1987), 210–23.

18. Obviously, this advice and counsel takes many forms in many different groups, varying with language, folkways, and styles of communication. One parent's advice may come in the form of constant reassurances and encouragement, while another may scold and demean the child when it misbehaves.

19. Cf. Susan Bordo, *Unbearable Weight: Feminism, Western Culture, and the Body* (Berkeley: University of California Press, 1993).

20. Brian Stableford, *Future Man* (New York: Crown Publishers, 1984), 13–15.

21. We have only to look at the tragic results of the introduction of ultrasound in India to see what thoughtless application of reproductive technologies can mean. Indian women are forced to abort their female fetuses despite the effect on the population and the women. Thus the very technology that was created to bring more of reproduction under the control of women came to be an instrument for the oppression of women—it is not the maldistribution of technology that is at issue, but the actual rearticulation of the purposes of that technology.

22. Cf. Glenn McGee, "Consumers, Land, and Food: In Search of Food Ethics," in A Bonanno, ed., *The Agricultural and Food Sector in the New Global Era* (New Delhi: Concept, 1993); Jack Doyle, *Altered Harvest: Agriculture, Genetics, and the Fate of the World's Food Supply* (New York: Viking, 1985); David Goodman, *From Farming to Biotechnology: A Theory of Agro-Industrial Development* (New York: Basil Blackwell, 1987); and House Hearings, "Field Testing Genetically-Engineered Organisms: Hearing Before the Subcommittee on Natural Resources, Agricultural Research, and Environment of the Committee on Science, Space, and Technology, U.S. House of Representatives, One Hundredth Congress, Second Session (Washington, D.C.: Governmental Publications Office, May 5, 1988).

23. For a full account of this controversy see Sheldon Krimsky, *Genetic Alchemy: The Social History of the Recombinant DNA Controversy* (Cambridge: MIT Press, 1982).

24. "The Construction of the Good," in J. McDermott, ed. *The Philosophy of John Dewey* (Chicago: University of Chicago Press, 1981), 577.
25. John Dewey (Quoted in McDermott, *The Philosophy of John Dewey*, 583).
26. Lewontin, *Biology as Ideology*.

Bibliography

Adams, M. 1990. *The Wellborn Science: Eugenics in Germany, France, Brazil, and Russia*, Oxford University Press, New York.

Agich, G. J. June 17, 1977. "Recombinant DNA Research and the Idea of Responsibility," paper presented to The Greater Delaware Philosophy and Technology Conference, University of Delaware.

Anderson, W. F. 1985. "Human Gene Therapy: Scientific and Ethical Considerations," *Journal of Medicine and Philosophy*, Volume 10, pp. 123–45.

Annas, G. 1992. *Gene Mapping: Using Law and Ethics as Guides*, Oxford University Press, New York.

Arkes, H. 1990. *Guaranteeing the Good Life: Medicine and the Return of Eugenics*, Eerdmans, New York.

Asch, Adrienne. 1995. "Parenthood and Embodiment: Reflections on Biology, Intentionality, and Autonomy," *Graven Images* 2: 229–36.

———. 1989 "Reproductive Technology and Disability" in S. Cohen and N. Taub, eds., *Reproductive Laws for the 1990s*, Humana, Clifton, N.J.

Auerbach, C. 1956. *Genetics in the Atomic Age*, Essential, New York.

Augustein, L. 1969. *Come Let Us Play God*, Macmillan, New York.

Bartels, D., LeRoy, B, and Caplan, A. 1993. *Prescribing Our Future: Ethical Challenges in Genetic Counseling*, de Gruyter, Hawthorne, N.Y.

Basen, G., et al. 1994. *The Social Construction of Choice and the New Reproductive Technologies*, Voyageur, Quebec.

Beadle, G. W. 1966.*The Language of Life: An Introduction to the Science of Genetics*, Doubleday, New York.

Bernard, K. 1990. *Genealogical Mathematics*, Bernard, Chicago.

Bichat, M. 1978. "Physiological Researches on Life and Death," in D. Robinson, ed., *Significant Contributions to the History of Responsibility*, Vol. 2, University Press of America, Lanham, Maryland.

Billings, P., and Hubbard, R. 1994. "Fragile X Tests," *geneWATCH* 9: 1.

Bishop, J. E. 1990. *Genome: The Story of the Most Astonishing Scientific Adventure of All Time*, Simon & Schuster, New York.

Blatt, J. R. 1988. *Prenatal Tests: What They Are, Their Benefits, and How to Decide Whether to Have Them or Not*, Vintage, New York.

Bock, K. 1980. *Human Nature and History: A Response to Sociobiology*, Columbia University Press, New York.

Bonnicksen, T. July 1992. "Genetics and the Moral Mission of Health Insurance," *Hastings Center Report*: S5.

Boone, C. K. August, 1988. "Bad Axioms in Genetic Engineering," *Hastings Center Report*: 10.

Borek, E. 1965. *The Code of Life*. Columbia University Press, New York.

Boswell, J. 1988. *Kindness of Strangers: The Abandonment of Children in Western Europe from Late Antiquity to the Renaissance*, Pantheon, New York.

Brand, C. 1994. "Criminal Portraits and the Psychology of Crime," *Nature* 368: 111–12.

Brennan, B. P. 1961. *The Ethics of William James*, Bookman, New York.

Brody, H. 1973. "The Systems View of Man: Implications for Medicine, Science, and Ethics," *Perspectives in Biology and Medicine* 17: 1–3.

Brown, G. 1994. "1994 Should See Firming Fundamentals," *Biotech* 12: 20–21.

Burnet, F. M. 1978. *Endurance of Life: The Implications of Genetics for Human Life*, Melbourne University Press, Australia.

Bynum, W. F. 1994. *Science and the Practice of Medicine in the Nineteenth Century*, Cambridge University Press, New York.

Callahann, J. C. 1988. *Ethical Issues in Professional Life*, Oxford University Press, New York.

Campbell, J. 1992. *The Community Reconstructs: The Meaning of Pragmatic Social Thought*, University of Illinois Press, Urbana.

———. 1981. "William James and the Ethics of Fulfillment," *Transactions of the Charles S. Peirce Society* 27: 224–40.

Caplan, A. 1995. *Moral Matters: Ethical Issues in Medicine and the Life Sciences*, Wiley, New York.

———. 1992. *If I Were a Rich Man, Could I Buy a Pancreas? And Other Essays on the Ethics of Health Care*, Indiana University Press, Bloomington.

Capron, A. M. September 1991. "Human Genome Research in an Interdependent World," *Kennedy Institute of Ethics Journal* 1: 247–51.

Cavalieri, L. F. 1981. *The Double-Edged Helix: Science in the Real World,* Columbia University Press, New York.

Chadwick, R. 1987. "Introduction," *Ethics, Reproduction, and Genetic Control*, Croom Helm, New York.

Chan, C. K. 1987. "Eugenics on the Rise: A Report from Singapore," in R. Chadwick, ed., *Ethics, Reproduction, and Genetic Control*, Croom Helm, New York.

Corcos, A. 1984. "Reproductive Hereditary Beliefs of the Hindus, Based on Their Sacred Books," *Journal of Heredity* 75: 152–54.

Cummings, M. R. 1988. *Human Heredity: Principles and Issues*, West, St. Paul, Minn.

Davis, B. D. 1991. *The Genetic Revolution: Scientific Prospects and Public Perceptions*, Johns Hopkins University Press, Baltimore.

Davis, J. 1990. *Mapping the Code: The Human Genome Project and the Choices of Modern Science*, Wiley, New York.

deGrouchy, J. 1984. *Clinical Atlas of Human Chromosomes*, Wiley, New York.

Dewey, J. 1991. *Logic: The Theory of Inquiry*, Southern Illinois University Press, Carbondale.

———. 1981. *The Philosophy of John Dewey,* Vols. I & II, J. McDermott, ed., University of Chicago Press, Chicago.

———. 1963. *Liberalism and Social Action*, Capricorn, New York.

———. 1963. *Philosophy and Civilization*, Capricorn, New York.

———. 1961. *Democracy and Education*, Macmillan, New York.

———. 1960. *The Quest for Certainty*, Capricorn, New York.

———. 1958. *Art as Experience*, Capricorn, New York.
———. 1958. *Experience and Nature*, Dover, New York.
———. 1957. *The Public and Its Problems*, Holt, New York.
———. 1930. *Human Nature and Conduct*, Holt, New York.
Dewey, J., and Tufts, J. H. 1932. *Ethics*, Holt, New York.
Dijksterhuis, E. J. 1976. *The Mechanization of the World Picture: Pythagoras to Newton*, Princeton University Press, Princeton.
Doyle, J. 1985. *Altered Harvest: Agriculture, Genetics, and the Fate of the World's Food Supply*, Viking, New York.
Dubos, R. 1990. *Mirage of Health: Utopias, Progress, and Biological Change*, Rutgers University Press, New Brunswick, N.J. (Reprint of 1959 ed.)
Dunn, L.C. 1965. *A Short History of Genetics*, McGraw-Hill, New York.
———. 1962. "Cross Currents in the History of Genetics," *American Journal of Human Genetics* 14: 1–13.
Edie, J. 1987. *William James and Phenomenology*, Indiana University Press, Bloomington.
Elias, S., and Annas, G. J. January 3, 1986. "Noncoital Reproduction," *Journal of the American Medical Association*: 67.
Ellos, W. 1984. "The Practice of Medical Ethics, A Structuralist Approach," *Theoretical Medicine* 5: 333–44.
Elmer-Dewitt, P. January 17, 1994. "The Genetic Revolution," *Time* 143: 46.
Elshtain, J. B. 1990. *Power Trips and Other Journeys: Essays in Feminism as Civic Discourse*, University of Wisconsin Press, Madison.
———. 1990. *Rebuilding the Nest: A New Commitment to the American Family*, Family Service America, Milwaukee, Wis.
———. ed. 1982. *The Family in Political Thought*, University of Massachusetts Press, Amherst.
Embree, L. 1988. "A Perspective on Scientific Technological Rationality, or How to Buy a Car" (mimeograph).
Engelhardt, H.T. 1980. "Ideology and Etiology," *Journal of Medicine and Philosophy*, 5: 256–63.
———. 1978. "Taking Risks: Some Background Issues in the Debate Concerning Recombinant DNA Research," *Southern California Law Review* 51: 114.
Erdrich, L. May 1993. "A Woman's Work," *Harper's Magazine*: 35–46.
FDA Consumer. April 1994. "Interleukin-2": 25–27.
Firestone, S. 1970. *Dialectic of Sex*, Morrow, New York.
Foucault, M. 1973. *The Birth of the Clinic*, trans. Sheridan Smith, Vintage, New York.
Fowler, J. 1981. *Stages of Faith: The Psychology of Human Development and the Quest for Meaning*, HarperCollins, New York.
Garber, S., et al. September 1993. "The Parent Trap: Perfections and Goals, *Sky*: 26–30.
Gavin, W. J. 1981. "Vagueness and Empathy, A Jamesian View," *Journal of Medicine and Philosophy* 6: 45–65.
Gaylin, W. 1990. *Being and Becoming Human*, Penguin, New York.
Genetic Engineering News. March 15, 1994. "Harvard and the Biotechnology Industry Unite": 1, 16.
Genius, S. J. 1993. "Public Attitudes in Edmonton Toward Assisted Reproductive Technologies," *Canadian Medical Association Journal* 149: 153–61.

Ginsburg, M. 1994. "Product Update," *Biotech* 12: 243–44, 358–59.
Glover, J. 1984. *What Sort of People Should There Be?* Penguin, New York.
Goodman, D. 1987. *From Farming to Biotechnology: A Theory of Agro-Industrial Development*, Basil Blackwell, New York.
Hamer, D. et al. 1993. "A Linkage Between DNA Markers on the X-Chromosome and Male Sexual Orientation," *Science* 261: 321–27.
Hardy, J. 1994. "ApoE, Amyloid, and Alzheimer's," *Science* 263: 454–55.
Harper's Magazine. May 1993. "Harper's Index" 286: 11.
Harris, J. 1992. *Wonderwoman and Superman: The Ethics of Human Biotechnology*, Oxford University Press, New York.
Hauerwas, S. 1985. "Suffering the Retarded: Should We Prevent Retardation?" Presented at Notre Dame University, South Bend, Ind.
Health Care Marketing Report. 1986. "Infant Sex Pre-selection Controversy Aids Marketeers" 89: 7.
Hickman, L. A. 1992. *John Dewey's Pragmatic Technology*, Indiana University Press, Bloomington.
Holtzman, N. A. 1990. "Research Discoveries and the Future of Screening," in B. M. Knoppers and C. M. Laberge, eds., *Genetic Screening: From Newborns to DNA Typing*, Excerpta Medica, New York, 354, 291–323.
———. 1989. *Proceed with Caution: Predicting Risks in the Recombinant DNA Era*, Johns Hopkins University Press, Baltimore.
———. 1988. *Nucleic Acid Probes in the Diagnosis of Human Genetic Disease*, Liss, New York.
Howard, T., and Rifkin, J. 1977. *Who Should Play God: The Artificial Creation of Life and What It Means for the Future of the Human Race*, Dell, New York.
Howell, J. D. 1995. *Technology in the Hospital: Transforming Patient Care in the Twentieth Century*, Johns Hopkins University Press, Baltimore.
Hu, S., et al. 1995. "Linkage Between Sexual Orientation and Chromosome Xq28 in Males But Not in Females," *Nature Genetics* 11: 248–56.
Human Genome News. January 1994. "ELSI CF Test Results.": 1–2.
Huxley, A. 1932. *Brave New World*, Harper & Row, New York.
Jacquard, A. 1984. *In Praise of Difference: Genetics and Human Affairs*, trans. Margaret Moriarty, Columbia University Press, New York.
James, W. 1981. *Pragmatism*, Bruce Kuklick, ed., Hackett, Indianapolis.
———. 1981. *The Principles of Psychology*, 3 vols., in *The Works of William James*, F. Burkhardt, ed., Harvard University Press, Cambridge.
———. 1958. *Talks to Teachers On Psychology; and To Students on Some of Life's Ideals*, Norton, New York.
———. 1956. *The Will to Believe and Other Essays in Popular Philosophy*, Dover, New York.
Jaroff, L. March 15, 1993. "Seeking a Godlike Power," *Time*: 56.
Jensen, A. R. 1978. "The Current Status of the I.Q. Controversy," *Australian Psychologist* 13: 7–27.
———. 1969. "How Much Can We Boost I.Q. and Scholastic Achievement?" *Harvard Educational Review*, 39: 6.
Jewson, N. D. 1976. "The Disappearance of the Sick Man from Medical Cosmology, 1770–1870," *Sociology* 10: 225–44.
Johanson, A. E. 1975. "The Will to Believe and the Ethics of Belief," *Transactions of the Charles S. Peirce Society* 11: 110–27.

Jonas, H. 1984. *The Imperative of Responsibility: In Search of an Ethics for the Technological Age*, University of Chicago Press, Chicago.

———. 1974. "Biological Engineering—A Preview," *Philosophical Essays from Ancient Creed to Technological Man*, Prentice-Hall, Englewood Cliffs, N.J.

Jones, D. G. 1984. *Brave New People: Ethical Issues at the Commencement of Life*, Intervarsity, Chicago.

Journal of the American Medical Association. 1994. "Alzheimer's Gene Reviewed" 271: 89–91.

Juengst, E. 1992. "Issues in Genetic: An Update," *Transcripts of the SHHV Annual Meeting*, audio recording, High Point Recordings, High Point, Missouri.

Kass, L. R. 1985. *Toward a More Natural Science: Biology and Human Affairs*, Free, New York.

———. 1971. "The New Biology: What Price Relieving Man's Estate," *Science* 174: 779–88.

Kass, N. December 1992. "Insurance for the Insurers: The Use of Genetic Tests," *Hastings Center Report*: 6–11.

Kegan, R. 1994. *In Over Our Heads: The Mental Demands of Modern Life*, Harvard University Press, Cambridge.

Keirns, C. 1996. "Review: The Science of Desire: The Search for the Gay Gene and the Biology of Behavior," *Journal of General Internal Medicine* 11: 444–45.

Kevles, D. J. 1986. *In the Name of Eugenics: Genetics and the Uses of Human Heredity*, University of California Press, Berkeley.

Kevles, D. J., and Hood, L., eds. 1992. *The Code of Codes: Scientific and Social Issues in the Human Genome Project*, Harvard University Press, Cambridge.

King, L. 1982. *Medical Thinking*, Princeton University Press, Princeton.

Kitcher, P. 1996. *The Lives to Come: The Genetic Revolution and Human Possibilities*, Simon & Schuster, New York.

Knoppers, B. M. and Laberge, C. M., eds. 1990. *Genetic Screening: From Newborns to DNA Typing*, Excerpta Medica, New York, 291–323.

Knowles, R. V. 1985. *Genetics, Society, and Decisions*, Merrill, Columbus, Ohio.

Kochanski, Z. 1971. "Conditions and Limitations of Prediction-Making in Biology" (mimeograph).

Korey, K. 1984. *The Essential Darwin*, Little, Brown, Boston.

Kranzberg, M. 1980. *Ethics in an Age of Pervasive Technology*, Westview, Boulder.

Krimsky, S. 1982. *Genetic Alchemy: The Social History of the Recombint DNA Controversy*, MIT Press, Cambridge.

Lachs, J. 1987. *Mind and Philosophers*, Vanderbilt University Press, Nashville.

Lain-Entralgo, P. 1969. *Doctor and Patient*, World University, New York.

Lebacqz, K. 1985. *Professional Ethics: Power and Paradox*, Abingdon, Nashville.

Lee, T. F. 1991. *The Human Genome Project: Cracking the Genetic Code of Life*, Plenum, New York.

Lewontin, R. C. April 7, 1994. "Women Versus the Biologists," *The New York Review of Books*, pp. 31–35.

———. 1991. *Biology as Ideology: The Doctrine of DNA*, Harper Perennial and Harper Torchbooks, New York.

———. 1982. *Human Diversity*, Scientific American Library, New York.

Lewontin, R. C. March 24, 1994. *Los Angeles Times* . "Genetic Danger" : 1A-12A.

Lewontin, R. C., S. Rose, and L. J. Kamin. 1984. *Not in Our Genes: Biology, Ideology, and Human Nature*, Random House, New York.

MacIntyre, A. 1977. "Can Medicine Dispense with a Theological Perspective on Human Nature?" in H. T. Engelhardt, ed., *Knowledge, Value, and Belief*, Institute of Society, Ethics, and the Life Sciences Press, Hastings-on-Hudson, N.Y.

Mayr, E. 1982. *Growth of Biological Thought: Diversity, Evolution, and Inheritance*, Harvard University Press, Cambridge.

McDermott, J. J. 1990. "Pragmatic Sensibility: The Morality of Experience," in J. DeMarco and R. Fox, eds. *New Directions in Ethics*, Routledge, New York.

———. 1986. *Streams of Experience: Reflections on the History and Philosophy of American Culture*, University of Massachusetts Press, Amherst.

McGee, D. B. 1992, "Making or Having Babies: A Christian Understanding of Responsible Parenthood," in W. Brackney, ed., *Faith, Life, and Witness: The Papers of the Study and Research Division of The Baptist World Alliance 1986–1990*, Samford University Press, Birmingham.

———. 1977, "The Questions of Modern Medicine," in H. Hollis, ed., *A Matter of Life and Death*, Broadman, Nashville.

McGee, G. January/February 1997. "A Few Deadly Sins of Genetic Enhancement," *Hastings Center Report.*

———. 1996. "American Literature and Science," *Society for Advancement of American Philosophy Report.*

———. 1996. "Frontiers in American Philosophy," *Transactions of the C. S. Peirce Society: A Quarterly Journal in American Philosophy.*

———. January 1996. "Who Keeps the Gender Gate: Ethical and Policy Issues in the Use of OB-GYNs as Family Practitioners," *American Journal of Managed Care* (with M. Arruda).

———. May 1996. "Waiting for Godot: Where Is the Philosophy of Nursing Today?" *Journal of Psychosocial Nursing.*

———. June 1996. "Disclosure vs. Confidentiality When Disaster Strikes," *Making the Rounds in Health, Faith, and Ethics.*

———. June 1996. "When Paternalism Runs Amok: Implications of Adolescent Sexual Activity for the Ability to Consent," *Politics and the Life Sciences* (with F. Burg).

———. July 1996. "American Literature and Science," *Society for Advancement of American Philosophy.*

———. August 1996. "Young Scientists Need to Feel a Personal Stake in Ethics," *Chronicle of Higher Education.*

———. Fall 1996. "The Human Genome Project and Reproductive Medicine: Ethical Issues," *OrGyn.*

———. Fall 1996. "Phronesis in Clinical Ethics," *Theoretical Medicine.*

———. 1995. "Designing Our Descendants? Outcome of an Interdisciplinary Conversation about Genetics," International Journal of Health Legislation.

———. 1994. "Method and Social Reconstruction: Dewey's *Logic: The Theory of Inquiry*," *Southern Journal of Philosophy* 32: 1, 107–20.

———. February 1994. "Pragmatism and the Human Genome Project," Distinguished Visiting Professor of Applied Ethics Lectures, California State University at Chico, (mimeograph).

McKusick, V. 1983. *Mendelian Inheritance in Man*, Johns Hopkins University Press, Baltimore.

McLoughlin, J. July 10, 1986. "Playing Russian Roulette with Employees," *Guardian*: 10.

Macklin, R. January 1991. "Artificial Means of Reproduction and Our Understanding of the Family," *Hastings Center Report*: 5–11.

Mann, C. C. 1994. "War of Words Continues in Violence Research," *Science* 263: 1375.

Mauron, A. Spring 1991. "Germ-Line Engineering: A Few European Voices," *Journal of Medicine and Philosophy*: 59–63.

Mestel, R. February 26, 1994. "What Triggers the Violence Within," *New Scientist* 263: 10, 31–33.

Murphy, E. A., and Chase, G. A. 1975. *Principles of Genetic Counseling*, Year Book Medical, Chicago.

Murray, T. December 1992. "Genetics and the Moral Mission of Health Insurance," *Hastings Center Report*: 12.

National Institutes of Health. 1982, *Guidelines for Research Involving Recombinant DNA Molecules*, Office of Publications of the Department of Health & Human Services, Washington D.C.

Nature. 1994. "The Genetics of Longevity and More Trinucleotide Repeats" 367: 201.

Nelkin, D. 1989. *Dangerous Dianostics: The Social Power of Biological Information*, Basic, New York.

Nelson, J. 1984. *Human Medicine: Ethical Perspectives on Today's Medical Issues*, Minneapolis University Press, Augsburg.

Nobel Conference. 1984. *The Manipulation of Life*, Harper & Row, New York.

Parker, L. S. 1994. "Bioethics for Human Geneticists: Models for Reasoning and Methods for Teaching," *American Journal of Human Genetics* 54: 124–36.

Parker, R. A., and Phillips, J. A. 1995. "Population Screening for Carrier Status," *American Journal of Medical Genetics* 54: 137–47.

Pattatucci, A., and Hamer, D. 1995. "Development and Familiality of Sexual Orientation in Females," *Behavior Genetics* 25:407–20.

Pierce, B.A. 1990. *The Family Genetic Sourcebook*, Wiley, New York.

Poole, R. 1993. "Evidence for a Homosexuality Gene," *Science* 261: 291–92.

Porter, R. 1993. "The Rise of Physical Examination," in W. F. Bynum and R. Porter, eds., *Medicine and the Five Senses,* Cambridge University Press, New York, 179–97.

Portman, A. 1990. *A Zoologist Looks at Humankind*, trans. by J. Schaefer, Columbia University Press, New York.

Rafter, N. H. 1988. *White Trash: The Eugenic Family Studies*, Northeastern University Press, Boston.

Ramsey, P. July 12, 1972. "Shall We 'Reproduce'?" *Journal of the American Medical Association* 220: 1484.

———. 1970. *Fabricated Man: The Ethics of Genetic Control*, Yale University Press, New Haven.

Rapp, R. 1987. "Moral Pioneers: Women, Men & Fetuses," *Women & Health* 13:101–16.

Re, R. N. 1986. *Bioburst: The Impact of Modern Technology on the Affairs of Man*, Louisiana State University Press, Baton Rouge.

Restak, R. 1975. *Pre-meditated Man: Bioethics and the Control of Future Human Life*, Viking, New York.

Rifkin, J. 1983. *Algeny: A New Word, a New World*, Penguin, New York.

Robertson, M. 1984. "Towards a Medical Eugenics?"*British Medical Journal* 288: 221–30.

Rosenberg, C. E. 1987. *The Care of Strangers: The Rise of America's Hospital System*, Basic, New York.

Rosenfeld, A. 1969. *The Second Genesis: The Coming Control of Life*, Vintage, New York.

Rosenthal, S. 1986. *Speculative Pragmatism*, Amherst Press, Amherst, Massachusetts.

Rostand, J. 1959. *Can Man Be Modified?: Predictions of Our Biological Future*, trans. Jonathan Griffin, Basic, New York.

Roth, J. K. 1969. *Freedom and the Moral Life*, Westminster, Philadelphia.

———. 1969. "Introduction," in J. K. Roth, ed., *The Moral Philosophy of William James*, Crowell, New York.

Rowland, R. 1992. *Living Laboratories: Women and Reproductive Technologies*, Indiana University Press, Bloomington.

———. 1987. "Motherhood, Patriarchal Power, Alienation, and the Issue of Choice in Sex Preselection," in R. Rowland, ed., *Man-Made Women: How New Reproductive Technologies Affect Women*, Indiana University Press, Bloomington.

Ryan, K. 1979. "Ethics and Pragmatism in Scientific Affairs," *Bioscience* 29: 35–37.

Science. 1994. "Is There an Addiction Gene?" 263: 176.

———. 1993. "New Piece in Alzheimer's Puzzle" 261: 828.

Shannon, T. A. 1985. *What Are They Saying About Genetic Engineering?* Paulist, Atlanta.

Shapiro, R. 1992. *The Human Blueprint: The Race to Unlock the Secrets of Our Genetic Script*, Bantam, New York.

Singer, M. G. 1985. "Moral Issues and Social Problems: The Moral Relevance of Moral Philosophy," *Philosophy* 60: 5–26.

Singer, P. 1985. *Making Babies: The New Science and Ethics of Conception*, Scribner's, New York.

Sleeper, R. W. 1986. *The Necessity of Pragmatism: John Dewey's Conception of Philosophy*, Yale University Press, New Haven.

Smith, J. E. 1978. *Purpose and Thought: The Meaning of Pragmatism*, Yale University Press, New Haven.

Spallone, P. 1992. *Generation Games: Genetic Engineering and the Future for Our Lives*, Temple University Press, Philadelphia.

Stableford, B. 1984. *Future Man*, Crown, New York.

Stein, H. 1982. *Ethics (and Other Liabilities): Trying to Live Right in an Amoral World*, St. Martin's, New York.

Stern, C. 1966. *The Origin of Genetics: A Mendel Sourcebook*, Freeman, San Francisco.

Stich, S. 1982. "The Genetic Adventure," *Report from the Center for Philosophy & Public Policy*: 10–11.

Stubbe, H. 1972. *History of Genetics: From Prehistoric Time to the Discovery of Mendel's Laws*, MIT Press, Cambridge.

Sullivan, M. 1986. "In What Sense Is Contemporary Medicine Dualistic?" *Culture, Medicine, and Psychiatry* 15: 10.

———. 1982. *Knowing and Healing: A Study of the Role of Self-Aware Activity in Medicine*, dissertation at Vanderbilt University, Nashville.

Suzuki, D.T., and Knudtson, P. 1988. *Genethics: The Ethics of Engineering Life*, Stoddart, New York.

Thompson, J. S. 1980. *Genetics in Medicine*, Saunders, New York.
Thompson, L. October 12, 1989. "Gene Screening May Alter Future," *Portland Evening Express*: 1, 6.
Tiley, N. A. 1983. *Discovering DNA: Meditations on Genetics and a History of the Science*, Van Nostrand Reinhold, New York.
Toombs, S. K. 1992. *The Meaning of Illness: A Phenomenological Account of Different Perspectives of Physician and Patient*, Kluwer, Dordrecht.
U.S. Department of Commerce. 1992. *Designing Genetic Information Policy: The Need for an Independent Policy Review of the Ethical, Legal, and Social Implications of the Human Genome Project: Sixteenth Report*, Government Publications, Washington D.C.
U.S. Department of Energy. 1992. *Human Genome 1991–92 Program Report*, Department of Commerce Technical Information Service, Washington, D.C.
U.S. Department of Health and Human Services and Department of Energy. 1990. *Understanding Our Genetic Inheritance: the U.S. Human Genome Project: the First Five Years FY 1991–1995*, Department of Energy Publications Office, Washington, D.C.
U.S. House of Representatives. 1988. "Field Testing Genetically-Engineered Organisms: Hearing Before the Subcommittee on Natural Resources, Agricultural Research, and Environment of the Committee on Science, Space, and Technology, U.S. House of Representatives, One Hundredth Congress, Second Session," Government Publications Office, Washington, D.C.
U.S. News and World Report. May 10, 1993. "Gene Therapy's Leap Beyond the Lab" 82: 82.
Utke, A. 1971. *Bio-Babel: Can We Survive the New Biology*, Knox, Atlanta.
Wallace, R. A. 1979. *The Genesis Factor*, Morrow, New York.
Watson, J. D. 1980. *The Double Helix: A Personal Account of the Discovery of the Structure of DNA*, (Critical Edition), G. Stint, ed., Norton, New York.
Wills, C. 1991. *Exons, Introns, and Talking Genes: The Science Behind the Human Genome Project*, Basic, New York.
Wingerson, L. 1990. *Mapping Our Genes: The Genome Project and the Futures of Medicine*, Dutton, New York.
Wright, R. July 9, 1990. "Achilles' Helix," *New Republic*: 21–31.
Zaner, R. M. Fall 1992. "New Genetics, New Ethics," *Delta Response: Tennessee Guild for Health Decisions* 1: 2.
———. 1988. *Ethics and the Clinical Encounter*, Prentice-Hall, Englewood Cliffs, N.J.
———. March 1984. "Genetic Engineering: A Moral Dilemma," Paper presented to "Workshop on Moral and Ethical Aspects of Genetic Engineering," Vanderbilt University, Nashville.
———. 1983. "The Clinical Body, The Medical Corpse: An Historical, Phenomenological Prologue" (mimeograph).
Zaner, R. M. and Wartofsky, K. 1980. "Editorial: Understanding and Explanation in Medicine," *Journal of Medicine and Philosophy* 5: 2–3.
Zilinskas, R. A. and Zimmerman, B. 1986. *The Gene Splicing Wars: Reflections on the Recombinant DNA Controversy*, Macmillan, New York.
Zimmerman, B. 1984. *Biofutures: Confronting the Genetic Era*, Plenum, New York.

Index

About the Author

Glenn McGee, the author of numerous articles and a prominent figure in the public conversation about genetics, is an assistant professor at the University of Pennsylvania Center for Bioethics and a Senior Fellow in the Wharton School's Leonard Davis Institute of Health Economics. He teaches in the philosophy department at Penn and coordinates research ethics for biomedical scientists-in-training at Penn. This is his first book.